Heidelberg
Science
Library

Gaylon S. Campbell

An Introduction to Environmental Biophysics

Springer-Verlag
New York
Heidelberg
Berlin

ylon S. Campbell
epartment of Agronomy and Soils
rogram in Biochemistry and Biophysics
Washington State University
Pullman, Washington 99163 / USA

Library of Congress Cataloging in Publication Data

Campbell, Gaylon S.
An introduction to environmental biophysics.
(Heidelberg science library)
Bibliography: p.
Includes index.
1. Biological physics. 2. Adaptation (Physiology) 3. Ecology.
I. Title. II. Series.
QH505.C34 574.1'91 76-43346

Printed in the United States of America.

ISBN 0-387-90228-7 Springer-Verlag New York

ISBN 3-540-90228-7 Springer-Verlag Berlin Heidelberg

Preface

The study of environmental biophysics probably began earlier in man's history than that of any other science. The study of organism–environment interaction provided a key to survival and progress. Systematic study of the science and recording of experimental results goes back many hundreds of years. Benjamin Franklin, the early American statesman, inventor, printer, and scientist studied conduction, evaporation, and radiation. One of his observations is as follows:

> My desk on which I now write, and the lock of my desk, are both exposed to the same temperature of the air, and have therefore the same degree of heat or cold; yet if I lay my hand successively on the wood and on the metal, the latter feels much the coldest, not that it is really so, but being a better conductor, it more readily than the wood takes away and draws into itself the fire that was in my skin.[1]

Franklin probably was not the first to discover this principle, and certainly was not the last. Modern researchers rediscover this principle frequently in their own work. It is sometimes surprising how slowly progress is made.

Progress in environmental biophysics, since the observations of Franklin and others, has been mainly in two areas: use of mathematical models to quantify rates of heat and mass transfer and use of the continuity equation that has led to energy budget analyses. In quantification of heat- and mass-transfer rates, environmental biophysicists have followed the lead of physics and engineering. There, theoretical and empir-

[1] From a letter to John Lining, written April 14, 1757. The entire letter, along with other scientific writings by Franklin, can be found in Reference [1.2].

ical models have been derived that can be applied to many of the transport problems encountered by the design engineer. These same models were applied to transport processes between living organisms and their surroundings.

This book is written with two objectives in mind. The first is to describe the physical microenvironment in which living organisms reside. The second is to present a simplified discussion of heat- and mass-transfer models and apply them to exchange processes between organisms and their surroundings. One might consider this a sort of engineering approach to environmental biology, since the intent is to teach the student to calculate actual transfer rates, rather than just study the principles involved. Numerical examples are presented to illustrate many of the principles, and problems are given at the end of each chapter to help the student develop skill in using the equations. Working of problems should be considered as essential to gaining an understanding of modern environmental biophysics as it is to any course in physics or engineering. The last four chapters of the book attempt to apply physical principles to exchange processes of living organisms. The intent was to indicate approaches that either could be or have been used to solve particular problems. The presentation was not intended to be exhaustive, and in many cases, assumptions made will severely limit the applicability of the solutions. It is hoped that the reader will find these examples helpful but will use the principles presented in the first part of the book to develop his own approaches to problems, using assumptions that fit the particular problem of interest.

Literature citations have been given at the end of each chapter to indicate sources of additional material and possibilities for further reading. Again, the citations were not meant to be exhaustive.

Many people contributed substantially to this book. I first became interested in environmental biophysics while working as an undergraduate in the laboratory of the late Sterling Taylor. Walter Gardner has contributed substantially to my understanding of the subject through comments and discussion, and provided editorial assistance on early chapters of the book. Marcel Fuchs taught me about light penetration in plant canopies, provided much helpful discussion on other aspects of the book, and read and commented on the entire manuscript. James King read Chapters 7 and 8 and made useful criticisms which helped the presentation. He and his students in zoology have been most helpful in providing discussion and questions which led to much of the material presented in Chapter 7. Students in my Environmental Biophysics classes have offered many helpful criticisms to make the presentation less ambiguous and, I hope, more understandable. Several authors and publishers gave permission to use figures, Karen Ricketts typed all versions of the manuscript, and my wife, Judy, edited the entire manuscript and offered the help and encouragement necessary to bring this project to completion. To all of these people, I am most grateful.

Pullman, 1977 G.S.C.

Contents

List of Symbols

A	area; amplitude of the diurnal temperature wave in soil
A_p	projected area on a plane perpendicular to the solar beam
A_h	projected area on a horizontal plane
a	absorptivity (subscripts: s, short wave; L, long wave); empirical coefficient; atmospheric attenuation coefficient
B_M	minimum metabolic rate for animals
b	empirical coefficient
C	fraction of sky covered by clouds; solution concentration
c	speed of light (3×10^8 m/s)
c_p	specific heat of air
c_s	specific heat of soil or other solid
c_b	body specific heat
D	diffusivity (subscripts: H, sensible heat; v, vapor; c, CO_2); damping depth
d	zero plane displacement; characteristic dimension; diameter of stomatal pore
E	water vapor flux density (subscripts: s, sweating; R, respiratory)
e	photon energy
F	mass flux density
$f(u)$	empirical wind function for Penman equation
G	soil heat flux density; conduction heat loss
$G(T)$	temperature function for photosynthesis
g	gravitational acceleration ($9.8 m/s^2$)

H	flux density of sensible heat
h_r	relative humidity
h	crop canopy height; Planck's constant (6.62×10^{-34} J s)
i_B	spectral emittance of a blackbody
J_w	liquid water flux density
K	crop canopy attenuation coefficient; rate constant for CO_2 fixation
K	eddy transport coefficient (subscripts: M, momentum; H, sensible heat; v, water vapor)
K_L	rate constant for light reaction of photosynthesis
k	hydraulic conductivity; von Karman constant (0.4); thermal conductivity
L	leaf area index
L^*	sunlit leaf area index
L	long-wave flux density (subscripts: i, incoming; oe, outgoing emitted)
l	length or distance
M	molecular weight; heat flux density to animal surface from metabolism (subscript: B, basal metabolic rate)
m	body mass; airmass number
n	number of moles
P	photosynthetic rate; air permeability of clothing; atmospheric pressure
P_M	photosynthetic rate at CO_2 saturation
P_{MLT}	photosynthetic rate at CO_2 and light saturation and optimum temperature
PAR	photosynthetically active radiation (0.4–0.7 μm)
p	water vapor pressure
q	rate of heat storage
R_{abs}	absorbed long- and short-wave radiation
R	gas constant (8.31 J K^{-1} mol^{-1}); plant respiration rate
R_n	net radiant flux density
r_H	sensible heat transport resistance (additional subscripts: a, air boundary layer; c, clothing or animal coat; t, animal tissue; b, coat plus tissue)
r_v	vapor transport resistance (additional subscripts: a, air boundary layer; c, clothing or coat; s, surface, stomate, or skin)
r_c	CO_2 transport resistance (additional subscripts: a, air boundary layer; s, stomate; m, mesophyll cell wall)
r_r	radiative transfer resistance ($\rho c_p/4\sigma T^3$)
r_e	parallel equivalent resistance of r_{Ha} and r_r

r	reflectivity
S	short-wave flux density (subscripts: i, incoming; t, total on horizontal surface; b, direct on horizontal surface; d, diffuse; p, direct perpendicular to beam)
S_{po}	solar constant (1.36 kW/m^2)
s	slope of the saturation vapor density curve; radius of cylinders or spheres for diffusion calculations
T	temperature (Kelvin or Celsius) (subscripts: a, ambient or air; b, body; w, wet bulb; d, dew point; o, ground or canopy surface temperature; e, equivalent blackbody temperature; ew, equivalent wet blackbody temperature)
t	time, transmissivity
t_o	time of solar noon
u	wind velocity in direction of mean wind; windspeed
u^*	friction velocity
u'	fluctuation of instantaneous wind about mean
u_c	windspeed in plant canopy
u_{ch}	windspeed at top of canopy
V	volume
v	lateral wind velocity
v'	fluctuation of lateral wind velocity
W	crop standing dry mass
w	vertical wind velocity
w'	fluctuation in vertical wind velocity
z	height above the soil surface
z_M, z_H, z_v	roughness parameters for momentum, heat, water vapor
α	surface albedo (short-wave reflectivity)
β	Bowen ratio
Γ	heat production per unit of oxygen consumed
γ	thermodynamic psychrometer constant ($\rho c_p/\lambda$)
γ^*	apparent psychrometer constant ($\gamma r_v/r_H$ or $\gamma r_v/r_e$)
δ	solar declination
ϵ	surface emissivity, dissipation rate for turbulent kinetic energy
ϵ_A	clear sky emissivity
ϵ_{AC}	cloudy sky emissivity
θ	zenith angle or angle from a surface normal
κ	thermal diffusivity of soil or other solids
λ	latent heat of vaporization; wavelength of light; latitude angle
υ	wavenumber or frequency of light; kinematic viscosity
ρ	air density

ρ_v, ρ_v'	water vapor density and saturation water vapor density (additional subscripts: a, ambient or air; s, evaporating surface; w, wet bulb)
ρ_s	soil or solid density
ρ_c	CO_2 density (subscripts: a, air; c, chloroplast)
ρ_b	animal body density
ρ_o	oxygen density in air (subscripts: e, expired; i, inspired; a, ambient)
σ	Stephan–Boltzmann constant (5.67×10^{-8} W $m^{-2}K^{-4}$)
τ	shear stress or momentum flux density; period of oscillation; time constant
ϕ	solar elevation angle
ϕ_M, ϕ_H	diabatic influence functions for momentum and heat
Φ	radiant flux density (subscript: B, blackbody)
Ψ	water potential (subscripts: g, gravitational; o, osmotic; m, matric; p, pressure)
ψ_M, ψ_H	diabatic profile correction functions for momentum and heat
ω	frequency ($= 2\pi/\tau$, where τ is the period of oscillation)
ζ	atmospheric stability index (+ is stable, − is unstable, 0 is neutral)
Gr	Grashof number
Nu	Nusselt number
Pr	Prandtl number (v/D_H)
Re	Reynolds number
Sc	Schmidt number (v/D_j)
Sh	Sherwood number
log	base 10 logarithm
ln	base e logarithm
exp	exponential ($\exp x = e^x$, where $e = 2.7183 \ldots$)

1 Introduction

The effects of the physical environment on behavior and life are such an intimate part of our everyday experience that one may wonder at the need to study them. Heat, cold, wind, and humidity have long been common terms in our language. A few simple experiments, however, can easily convince us that our common expressions lack precision in explaining interactions with our environment, and in many cases we mislead ourselves or unnecessarily complicate the picture by not understanding the physical principles involved in environment–organism interactions.

One experiment can be tried the next time you step from a warm shower on a cold morning. By standing on the tile floor and then on the bath mat, try to estimate the relative temperature of each. Which is colder? Our senses tell us that the tile is colder, but if we were to make a measurement with a thermometer we would likely find them to be at the same temperature. What, then, did we sense? With little thought we will conclude that we sensed heat flux. Because of its higher thermal conductivity, the tile caused more rapid loss of heat from our feet than did the bath mat, and we registered that as a colder temperature. Careful consideration will indicate that essentially every interaction we have with our surroundings involves energy or mass exchange. Sight is possible because emitted or reflected photons from our surroundings enter the eye and cause photochemical reactions at the retina. Hearing results from the absorption of acoustic energy from our surroundings. Smell involves the flux of gasses and aerosols to the olfactory sensors. We could list numerous other sensations—such as sunburn, heat stress, cold stress—and each involves the flux of something to or from the organism. We can express the steady-state exchange of most forms of matter and energy between organisms and their surroundings as

$$\text{Flux} = \frac{C_s - C_a}{r} \tag{1.1}$$

where C_s is the concentration at the organism exchange surface, C_a is the ambient concentration, and r is the exchange resistance. As we already noted, we sense fluxes but we generally interpret them in terms of ambient concentrations. Even if we maintained the concentration at the organism constant (generally not the case) our judgment about ambient concentration would always be tempered by the magnitude of the exchange resistance. The shower experiment illustrates this nicely. The bath mat resistance to heat transfer was higher than that of the tile, so we judged the bath mat temperature to be higher.

Microenvironments

Microenvironment is an intimate part of our everyday life, but we seldom stop to think of it. Our homes, our beds, our cars, the sheltered side of a building, the shade of a tree, an animal's burrow are all examples of microenvironments. The ''weather'' in these places cannot usually be described by measured and reported weather data. The air temperature may be 10°C and the wind 5 m/s, but a bug, sitting in an animal track sheltered from the wind and exposed to solar radiation may be at a comfortable 25°C.

It is the microenvironment that is important when considering organism energy exchange, but descriptions of microclimate are often complicated because the organism influences its microclimate and because microclimates are extremely variable over short distances. Specialized instruments are necessary to measure relevant environmental variables. Variables of concern may be temperature, atmospheric moisture, radiant energy flux density, wind, oxygen and CO_2 concentrations, temperature and thermal conductivity of the substrate (floor, ground, etc.), and possibly spectral distribution of radiation. Other microenvironmental variables may be measured for special studies.

We will first concern ourselves with a study of the environmental variables—namely, temperature, humidity, wind, and radiation. We will then discuss energy and mass exchange, the fundamental link between organisms and their surroundings. Next we will apply the principles of energy and mass exchange to a few selected problems in plant, animal, and human environmental biophysics. Finally, we consider some problems in radiation, heat, and water vapor exchange for vegetated surfaces such as crops or forests.

Energy Exchange

The fundamental concept behind all of biophysical ecology is energy exchange. Energy may be exchanged as stored chemical energy, heat energy, or mechanical energy. Our attention will be focused primarily on the transport of heat energy.

Four modes of heat transfer are generally recognized. These are convection, conduction, radiative exchange, and latent heat transfer. These modes of energy transport are recognized in our common language when we talk of the ''hot'' sun (radiative exchange) or the ''cold'' floor tile (conduction), the ''chilling'' wind (convection), or the ''stifling'' humidity (reduced latent heat loss). An understanding of the principles behind each of these processes will provide the background we need to determine the physical suitability of a given environment for a particular organism.

The total heat content of a substance is proportional to the total random kinetic energy of its molecules. Heat can flow from one substance to another if the average kinetic energies of the molecules in the two substances are different. Temperature is a measure of the average random kinetic energy of the molecules in a substance. If two substances at different temperatures are in contact with each other, heat is transferred from the high-temperature substance to the low by conduction, a direct molecular interaction. If you touch a hot stove, your hand is heated by conduction.

Heat transport by a moving fluid is called convection. The heat is first transferred to the fluid by conduction, but the fluid motion carries the heat away. Most home heating systems rely on convection to heat the air and walls of the house.

Unlike convection and conduction, radiative exchange requires no intervening molecules to transfer heat from one surface to another. A surface radiates energy at a rate proportional to the fourth power of its absolute temperature. Both the sun and the earth emit radiation, but because the sun is at a higher temperature the emitted radiant flux density is much higher for the sun's surface than for the earth's surface. Much of the heat you receive from a campfire or a stove may be by radiation, and your comfort in a room is often more dependent on the amount of radiation you receive from the walls than on the air temperature.

To change from a liquid to a gaseous state at 20 °C, water must absorb about 2450 joules per gram (the latent heat of vaporization), almost 600 times the energy required to raise the temperature of one gram of water by one degree. Evaporation of water from an organism can therefore be a very effective mode of energy transfer. Almost everyone has had the experience of stepping out of a swimming pool on a hot day and feeling quite cold until the water dries from his skin.

Mass and Momentum Transport

Organisms in natural environments are subject to forces of wind or water on them, and rely on mass transport to exchange oxygen and carbon dioxide. The force of wind or water on an organism is a manifestation of the transport of momentum from the fluid to the organism. Transport of momentum, oxygen, and carbon dioxide in fluids follows principles similar to those developed for convective heat transfer. We can therefore learn just one set of principles and apply it to all three areas.

Applications

From the examples already given, it is quite obvious that environmental biophysics can be applied to a broad spectrum of problems. Fairly complete evaluations already exist for some problems, though much work remains to be done. Analysis of human comfort and survival in hot and cold climates requires a good understanding of the principles we will discuss. Preferred climates, survival, and food requirements of domestic and wild animals can also be considered. Plant adaptations in natural systems can be understood, and optimum plant types and growing conditions in agriculture and forestry can be selected through proper application of these principles. Even the successful architectural design of a building, which makes maximum use of solar heat and takes into account wind and other climatologic variables, requires an understanding of this subject.

Table 1.1 Examples of derived SI units and their symbols

Quantity	Name	Symbol	SI base units	Derived units
area	square meter	—	m^2	—
volume	cubic meter	—	m^3	—
velocity	meter per second	—	$m\ s^{-1}$	—
density	kilogram per cubic meter	—	$kg\ m^{-3}$	—
force	newton	N	$m\ kg\ s^{-2}$	—
pressure	pascal	Pa	$m^{-1}kg\ s^{-2}$	$N\ m^{-2}$
energy	joule	J	$m^2\ kg\ s^{-2}$	$N\ m$
power	watt	W	$m^2\ kg\ s^{-3}$	$J\ s^{-1}$
heat flux density	watt per square meter	—	$kg\ s^{-3}$	$W\ m^{-2}$
specific heat capacity	joule per kilogram kelvin	—	$m^2s^{-2}K^{-1}$	$J\ kg^{-1}K^{-1}$

Negative exponents are used to indicate units in the denominator

As we study environmental biophysics, we will find that animals and people from ''primitive'' cultures have a far better understanding of the application of its principles than we do. This application often makes the difference between life and death for them, whereas for us it may just mean a minor annoyance or an increased fuel bill.

Units

Units consistent with the Systéme International (SI) will be used in this book. Use of these units may seem a bit awkward at first, but will ultimately result in considerable savings in thought, time, and paper, since conversions to reconcile units will not be necessary each time a calculation is made. The SI base units and their accepted symbols are the meter (m) for length, the kilogram (kg) for mass, the second (s) for time, and the Kelvin (K) for thermodynamic temperature. Some derived units are given in Table 1.1. Additional derived units can be found in Reference [1.1].[1]

To make the numbers used with these units convenient, prefixes are attached indicating decimal multiples of the units. Accepted prefixes, symbols, and multiples are shown in Table 1.2. The use of prefix steps smaller than 10^3 is discouraged. Prefixes can be used with base units or derived units, but may not be used on units in the denominator of a derived unit (e.g., g/m^3 or $\mu g/m^3$, but not $\mu g/cm^3$).

The Celsius temperature scale is more convenient for some biophysical problems than the thermodynamic (Kelvin) scale. We will use both. By definition °C = K−273.15. Some useful factors for converting to SI units can be found in Appendix A.4.

[1] Numbers in brackets refer to references, which are listed at the end of each chapter.

Table 1.2 Accepted SI prefixes and symbols for multiples and submultiples of units

Multiple	Prefix	Symbol	Submultiple	Prefix	Symbol
10^{12}	tera	T	10^{-1}	deci	d
10^{9}	giga	G	10^{-2}	centi	c
10^{6}	mega	M	10^{-3}	milli	m
10^{3}	kilo	k	10^{-6}	micro	μ
10^{2}	hecto	h	10^{-9}	nano	n
10^{1}	deka	da	10^{-12}	pico	p
			10^{-15}	femto	f
			10^{-18}	atto	a

Transport Laws

The rate of transport of mass or energy is usually expressed as the product of a proportionality factor and a driving force. The transport laws with which we are most familiar are the following:

Newton's law of viscosity:

$$\tau = \mu \frac{du}{dx} \quad (1.2)$$

Fick's diffusion law:

$$F_j = -D_j \frac{d\rho_j}{dx} \quad (1.3)$$

Fourier's heat-transfer law:

$$H = -k \frac{dT}{dx} \quad (1.4)$$

where τ is the shear stress (N/m^2) between layers of a moving fluid with dynamic viscosity, μ, and velocity gradient, du/dx; F_j is the flux density (g m^{-2}s^{-1}) of a diffusing substance with a molecular diffusivity, D_j (m^2/s) and a concentration (g/m^3) gradient of $d\rho_j/dx$; and H is the heat flux density (W/m^2) in a substance with thermal conductivity, k, and temperature gradient, dT/dx. Each of these equations expresses a linear relationship between a flux density and a driving force. The negative sign in Equations 1.3 and 1.4 indicates that the flux is in the positive direction when the gradient is negative.

For mathematical manipulations it would be convenient to have all of the transport equations in the same form. It is a simple matter to express Equations 1.2 and 1.4 in the form of diffusion equations with the flux density described as the product of a diffusivity and a concentration gradient. Thus, Equation 1.2 becomes

$$\tau = v \frac{d(\rho u)}{dx} \; . \quad (1.5)$$

If ρ is the fluid density, ρu is a concentration of momentum and v ($= \mu/\rho$) is a momentum diffusivity called the kinematic viscosity.

Equation 1.4 is converted to a diffusion equation by changing the temperature to a heat concentration through multiplication by the volumetric heat capacity of air, ρc_p (the value of ρc_p will be taken as 1200 J m^{-3}K^{-1} throughout this book). The thermal diffusivity is therefore $D_H = k/\rho c_p$, and the heat transfer equation becomes

$$H = D_H \frac{d(\rho c_p T)}{dx} \quad . \tag{1.6}$$

There are several advantages to expressing all of the transport equations in diffusion equation form. The mathematical manipulations are the same for all the transport processes, and the units (m^2/s) are the same for all of the diffusivities. The diffusivities are roughly the same size, and they all have similar temperature and pressure dependence (which can be derived from kinetic theory). For many conditions of interest in environmental biophysics, the diffusivities are constant multiples of each other so that if one is known, the others can be easily found.

It is important to recognize that conversion of the heat transfer equation to diffusion form is just a mathematical manipulation and does not imply that a gradient in heat concentration is the driving force for heat flow. Multiplication of the temperature by the constant ρc_p is a means of converting a temperature, which *is* the driving force for heat flow, into a heat concentration. It is relatively unimportant what volumetric specific heat is used for this conversion, providing one is consistent in defining D_H. Through most of this book we will use the volumetric specific heat of air to convert temperature into heat concentration, and define D_H as $k/\rho c_p$.

In addition to the transport laws previously mentioned, a fourth law is in common use that describes the flow of current in an electrical circuit. This is Ohm's law, which states that the current flowing in a conductor is directly proportional to the applied voltage and inversely proportional to the electrical resistance of the conductor. This law is different from the other laws in that it applies to a macroscopic system. The applied voltage is measured across an entire conductor, not over an infinitesimal increment as is indicated by the differentials in Equations 1.2 to 1.6. The resistance of the conductor is a function of size and shape as well as basic material properties.

In environmental biophysics, our problems are similar to the circuit problem. It is usually possible to specify concentrations at the organism surface and in the surroundings some distance from the organism, but it is usually impossible to measure gradients on a microscopic scale, as would be needed for the transport equations as we have defined them so far. We therefore write the transport equations, by analogy with Ohm's law, in an integrated or macroscopic form such as Equation 1.1. The concentrations are specified at the organism surface and in the surroundings, and the transport resistance is defined as the concentration difference divided by the flux density. Chapter 6 will deal in detail with integration of the transport equations to determine resistance values from basic fluid properties and system geometries. Chapters 2 and 3 will describe the ambient environment of organisms and methods for estimating surface concentrations.

There are several good reasons for writing the transport equations in a form analogous to Ohm's law and using resistances rather than conductances for transport coefficients.

1. It is often convenient to represent transport processes between an organism and its environment by an equivalent electrical circuit. Concentrations are easily expressed as voltages, flux densities as currents, and transport resistances as electrical resistances.
2. The units of resistance are simple (s/m); and for many transport processes simple proportional relationships exist between resistances for various components.
3. Circuit analysis equations for series and parallel resistors can be applied to transport problems.
4. The resistance concept is well established in the plant physiology literature for describing gas exchange between leaves and the atmosphere.

References

1.1 Page, C. H. and P. Vigoureux (1974) *The International System of Units*. Nat. Bur. Stand. Spec. Publ. 330 U.S. Govt. Printing Office, Washington, D.C.

1.2 Seeger, R. J. (1973) *Benjamin Franklin, New World Physicist*. New York: Pergamon Press.

2 Temperature

Rates of biochemical reactions within an organism are strongly dependent on its temperature. The rates of reactions may be doubled or tripled for each 10 °C increase in temperature within a certain range. Temperatures above or below critical values may result in denaturation of enzymes and death of the organism.

A living organism is seldom at thermal equilibrium with its environment, so the environmental temperature is only one of the factors determining organism temperature. Other influences are fluxes of radiant and latent heat to and from the organism, heat storage, and resistance to sensible heat transfer between the organism and its surroundings. Sensible heat flux density is proportional to the temperature difference between the organism and its surroundings, and inversely proportional to the resistance to heat transfer.

Following Equation 1.1 we can write

$$H = \rho c_p \frac{(T_s - T_a)}{r_H} \tag{2.1}$$

where H is sensible heat flux density, T_a and T_s are ambient environment and organism exchange surface temperatures, ρc_p is the volumetric heat capacity of air (1200 J $m^{-3}K^{-1}$), and r_H is the resistance to heat transfer. Note that when $T_s > T_a$, H is positive, indicating flux of heat away from the organism.

Even though environmental temperature is not the only factor determining organism temperature, it is nevertheless important. In this chapter we want to look at environmental temperature in the biosphere and talk about reasons for its observed characteristics.

Typical Behavior of Atmospheric and Soil Temperature

If we were to measure average temperatures at various heights above and below the ground and plot these as a function of height, we would obtain graphs similar to Figure 2.1. Radiant energy input and loss is at the soil or vegetation surface. As the surface heats, heat is transferred away from the surface by convection to the air layers above and by conduction to the soil beneath the surface. We note that the temperature extremes occur at the surface, where temperatures may be 5–10 °C different from temperatures measured 2 m above the surface. Thus temperatures measured for standard meteorological observations often are not relevant to microclimate studies.

A typical temperature vs. time curve for air is shown in Figure 2.2. Similar curves would be obtained for temperatures measured a few centimeters below the soil surface. The fact that the temperature maximum lags the time of maximum radiant energy input is significant. This type of lag is typical of any system with storage and resistance. The closer one makes the measurement to the exchange surface the smaller will be the lag or phase shift, because the heat storage between the exchange surface and the point of measurement becomes small, but maximum temperature lags maximum heat input even at the exchange surface. The principles involved are illustrated by reconsidering the tile floor experiment from Chapter

FIGURE 2.1. Temperature profiles (hypothetical) just above and below the soil surface on a clear, calm day.

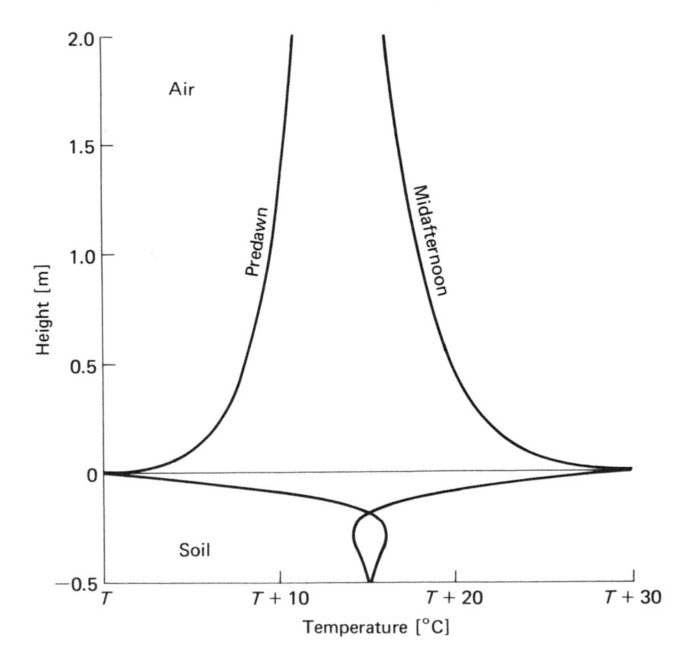

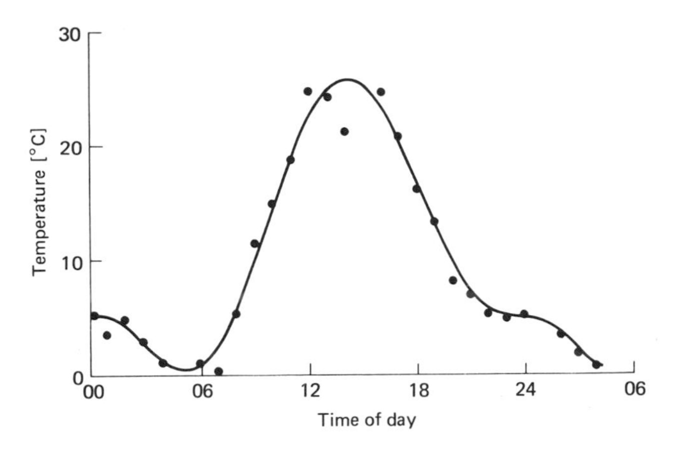

FIGURE 2.2. Air temperature 3 m above the ground on a clear September day in Pullman. The solid line is a polynomial fit to the data points.

1. The floor feels coldest (maximum heat flux to the floor) when your foot just comes in contact with it, but the floor reaches maximum temperature at a later time when heat flux is much lower.

The amplitude of the diurnal temperature wave becomes smaller with distance from the exchange surface. For the soil, this is because heat is stored in each succeeding layer so less heat is passed on to the next layer. At a depth of 50 cm or so, the diurnal temperature fluctuation in the soil is hardly measurable any more (Figure 2.1).

The diurnal temperature wave penetrates much farther in the atmosphere than in the soil because heat transfer in the atmosphere is by eddy motion, or transport of "blobs" of hot or cold air over relatively long vertical distances, rather than by molecular motion. Within the first few meters of the earth's atmosphere, the vertical distance over which eddies can transport heat is directly proportional to their height above the soil surface. The larger the transport distance, the more effective eddies are in transporting heat, so the air becomes increasingly well mixed as one moves away from the earth's surface. This mixing evens out the temperature differences between layers. This is the reason for the shape of the air temperature profiles in Figure 2.1. They are steep close to the surface because heat is transported only short distances by the small eddies there. Farther from the surface the eddies are bigger, so the change in temperature with height (temperature gradient) becomes much smaller.

Microenvironment Temperatures

With some understanding of the general behavior of temperature in nature, we can now focus on some details. As has been pointed out, the temperatures of microenvironments inhabited by living organisms are generally not the same as air temperature measured in a standard weather shelter and reported by the National Weather Service or the local TV station. Here we will attempt to bring out some of the differences and indicate possible techniques that can be used to estimate microenvironment temperature from standard observations.

Since heat transfer in air is mainly by convection, or transport of "blobs" of hot or cold air, we might expect the temperature at any point in the air to be quite different from the mean air temperature measured with a mercury-glass thermometer. Figure 2.3 shows the temperature variation that is observed with a 25-μm-diameter thermocouple thermometer on a hot, clear day. Measurements with smaller thermometers show even more detail [2.3]. The relatively smooth baseline with jagged interruptions indicates a suspension of hot blobs in relatively uniformly cold air. One might expect this since cold, well-mixed air is subsiding, being heated at the soil surface, and breaking away from the surface as convective bubbles when local heating is sufficient.

Temperature fluctuations in the atmosphere give rise to a number of interesting natural phenomena. The twinkling of stars and the scintillation of terrestrial light sources at night result from temperature-induced refractive index fluctuations in the air. The hot and cold air blobs form a diffraction pattern which is swept along with the wind. If you look at a city on a clear night from some distance, you can see the drift of the scintillation pattern.

So-called "heat waves" often seen on clear days are also refractive index fluctuations. The drift of heat waves can be seen, and wind direction and speed can sometimes be esti-

FIGURE 2.3. Air temperature 2 m above a dry desert surface near midday measured with a 25-μm-diameter thermocouple.

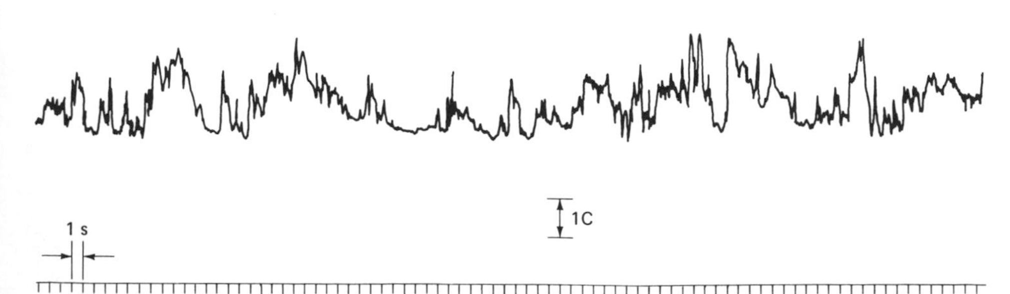

mated from the drift velocity. This phenomenon has been used to measure windspeed [2.4].

Air temperatures are often specified with a precision of 0.5 to 0.1 °C. From Figure 2.3 we see that temperature would need to be averaged over a relatively long time period to make this level of precision meaningful. Averages over 15 to 30 minutes are generally used. Figures 2.1 and 2.2 show the behavior of such long-term temperature averages.

The theory of turbulent transport, which we will study in Chapter 4, specifies the shape of the temperature profile over a uniform surface with steady-state conditions. The temperature profile equation is

$$\overline{T} = T_o - \frac{H}{\rho c_p k u^*} \ln \frac{z + z_H - d}{z_H} \tag{2.2}$$

where $\overline{T}$ is the mean air temperature at height z, T_o is the temperature at the exchange surface ($z = d$), z_H is a roughness parameter for heat transfer, H is the sensible heat flux from the surface to the air, ρ and c_p are air density and specific heat ($\rho c_p = 1200$ J m^{-3}K^{-1}), k is the von Karman constant (0.4), and u^* is the friction velocity, a windspeed and surface roughness parameter. The reference level from which z is measured is always somewhat arbitrary, and the correction factor, d, called the zero-plane displacement, is used to adjust for this. For a flat, smooth surface, $d = 0$.

We will not go into detail now on finding u^*, z_H, and H. We only want to use Equation 2.2 to point out some things about the temperature profile. The important points are:

1. Near the surface the temperature profile is logarithmic (e.g., a plot of $\ln(z + z_H - d)$ vs. $\overline{T}$ is a straight line).
2. Temperature increases with height when H is negative (heat flux toward the surface) and decreases with height when H is positive. During the day, sensible heat flux is generally away from the surface so T decreases with height.
3. The temperature gradient at a particular height increases in magnitude as the magnitude of H increases, and decreases as wind or turbulence increase.

Figure 2.4 shows a typical daytime temperature profile plotted linearly and as a function of $\ln(z - d + z_H)$. The logarithmic plot produces a straight line which can be extrapolated to $\ln z_H$ to determine the mean surface temperature. Measurements of mean air temperature at two or more heights can be plotted logarithmically, as is shown in Figure 2.4, and used

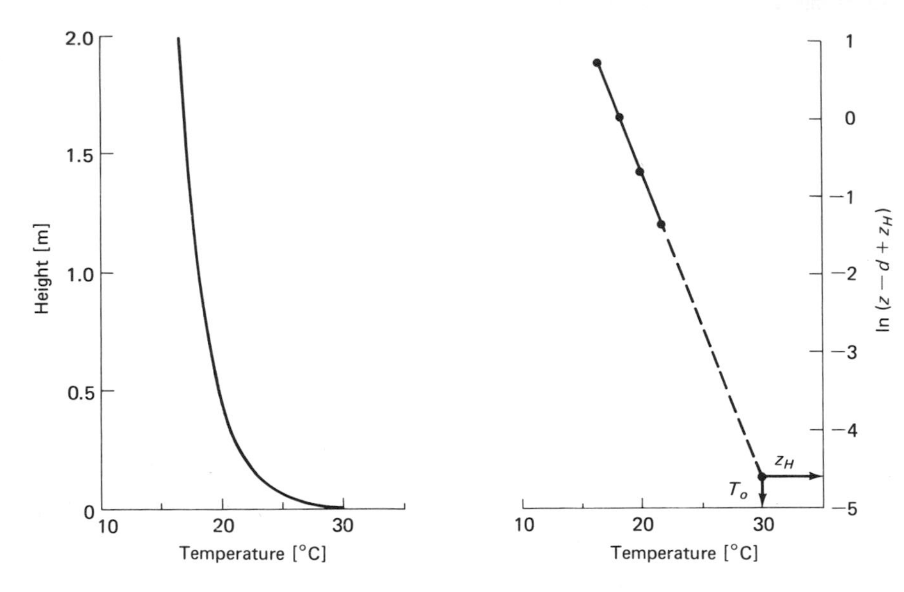

FIGURE 2.4. Typical daytime temperature profile plotted as a function of height (left) and logarithm of height (right). The log plot shows the extrapolation of the measured profile to $z - d = z_H$ to determine the surface temperature.

to determine the mean temperature at other heights above the exchange surface ($z = d$). Independent estimates of z_H and d are required, but these are often simple functions of canopy height, as we will see in Chapter 4. This approach, of course, ignores horizontal temperature variability, which is sometimes sizable [2.2]. One also needs to remember that the estimates are only for mean temperature, but they can still be useful for many studies of organism environments.

Soil Temperature

A mathematical model of soil temperature must include both conduction from one soil depth to the next and heat storage within soil at a given depth. If we consider a soil slab of unit area, with depth Δz, as is shown in Figure 2.5, all of the heat flowing in at a depth z must either flow out at depth $z + \Delta z$ or be stored in the soil. The Fourier law can be used to describe the heat flux density at z and $z + \Delta z$:

$$G = -k\frac{dT}{dz} \tag{2.3}$$

where k is the thermal conductivity of the soil and dT/dz is the temperature gradient. The rate of heat storage per unit area in the thin layer of soil is the heat capacity of the layer multiplied

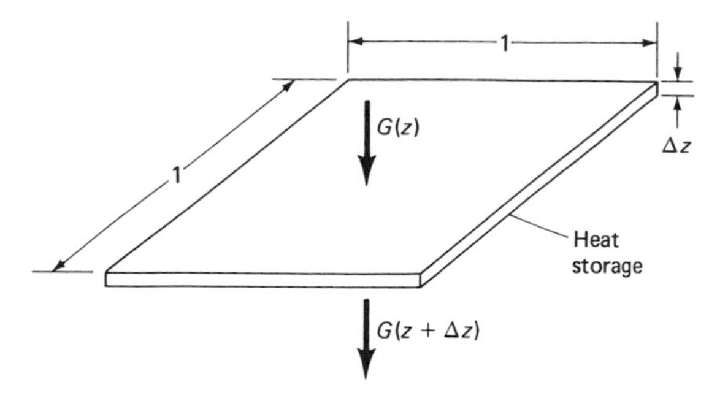

FIGURE 2.5. A thin slab of soil of unit area showing fluxes into and out of the system and storage within the system.

by the rate of change of temperature or

$$\text{Heat storage rate} = -\rho_s c_s \, \Delta z \frac{dT}{dt} \tag{2.4}$$

where ρ_s and c_s are the density and specific heat of soil, and t is time. Now, since the heat storage rate equals the difference between the flux in at z and the flux out at $z + \Delta z$, we can use Equations 2.3 and 2.4 to write

$$k \left.\frac{dT}{dz}\right|_{z+\Delta z} - k \left.\frac{dT}{dz}\right|_{z} = \rho_s c_s \, \Delta z \frac{dT}{dt} \tag{2.5}$$

where the thermal conductivity, k, is assumed to remain constant with depth. The vertical bar and subscript after the derivatives indicates the depth at which the gradient is measured. Rearranging this equation gives the rate of change of temperature with time:

$$\frac{dT}{dt} = \frac{\kappa}{\Delta z} \left\{ \left.\frac{dT}{dz}\right|_{z+\Delta z} - \left.\frac{dT}{dz}\right|_{z} \right\} \tag{2.6}$$

where $\kappa \, (= k/\rho_s c_s)$ is the thermal diffusivity of the soil. Equation 2.6 says that if the difference in temperature *gradient* is large between two ajacent levels in the soil, the rate of change of temperature in that layer will be large.

Equation 2.6 is not very useful unless we can integrate it. Integration of equations like Equation 2.6 is generally extremely difficult because they involve more than one variable. A solution may be obtained by assuming that the soil is uniform, and infinitely deep and that the temperature at the soil surface is given by

$$T\,(0,t) = \overline{T} + A(0) \sin \omega t \tag{2.7}$$

where $\bar{T}$ is the average surface temperature, $A(0)$ is the amplitude of the surface temperature fluctuation [maximum daytime temperature is $\bar{T} + A(0)$ and minimum night temperature is $\bar{T} - A(0)$], and ω is the angular frequency of the oscillation given by $2\pi/\tau$, where τ is the period of the oscillation (e.g., 24 hr for a daily cycle, 365 days for an annual cycle). Time is taken as zero when $T = \bar{T}$ and is increasing. If the temperature at the surface can be approximated by Equation 2.7, then the temperature at any depth and time is given by

$$T(z,t) = \bar{T} + A(0)e^{-z/D} \sin(\omega t - z/D) \tag{2.8}$$

where D $[=(2\kappa/\omega)^{1/2}]$ is called the damping depth. When $z = D$, $A(z) = A(0)e^{-1} = 0.37\,A(0)$, or the temperature fluctuation has been reduced to 37 percent of its surface value. At $z = 2D$, $A(z)$ is $A(0)e^{-2} = 0.14$, or only 14 percent of the surface value. Equation 2.8 also says that the time of the maximum and minimum temperature will be shifted with depth. Figure 2.6 shows $T(z,t)$ from Equation 2.8 over a two-day cycle. The reduction in amplitude with depth is evident, as is the shift of maxima to later times with increasing depth.

In order to find the damping depth, D, it is necessary to know κ, the thermal diffusivity of the soil. The thermal diffusivity is the ratio of thermal conductivity to volumetric specific heat, both of which change with soil water content and composition of the solid phase. Figure 2.7 gives values for κ in typical soils as a function of water content.

As an example of uses that can be made of this information, we will determine diurnal and annual damping depths for a typical moist soil. At intermediate water contents and densities, soil thermal diffusivities are around 0.5 mm^2/s (Figure 2.7). The diurnal (τ = 86400 s, $\omega = 7.3 \times 10^{-5}$ s^{-1}) damping depth for this value of κ is

$$D_{\text{diurnal}} = \left(\frac{2\kappa}{\omega}\right)^{1/2} = \left(\frac{2 \times 5 \times 10^{-7}\ \text{m}^2/\text{s}}{7.3 \times 10^{-5}\ \text{s}^{-1}}\right)^{1/2} = 0.12\ \text{m}$$

and the annual ($\tau = 3.15 \times 10^7$ s, $\omega = 2 \times 10^{-7}$ s^{-1}) damping depth is

$$D_{\text{annual}} = \left(\frac{2 \times 5 \times 10^{-7}\ \text{m}^2/\text{s}}{2 \times 10^{-7}\ \text{s}^{-1}}\right)^{1/2} = 2.24\ \text{m}.$$

We might therefore expect the temperature at 30 cm or so to be about equal to average daily temperature and the temperature at 6 or 7 m to be about equal to average annual temperature.

The heat flux into or out of the soil can be found by dif-

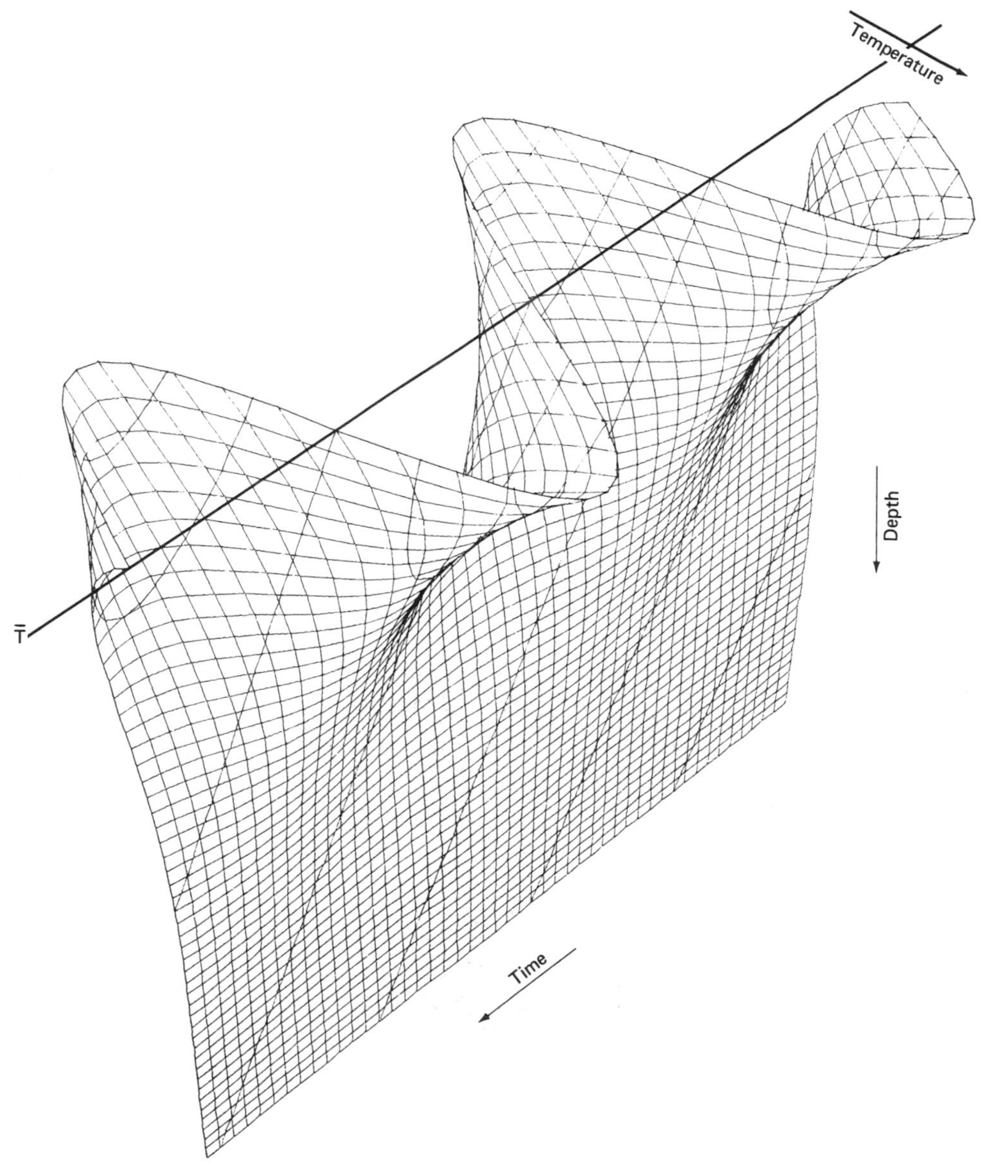

FIGURE 2.6. Three-dimensional plot of soil temperature as a function of time and depth showing the attenuation of the temperature wave with depth and the shift of maxima and minima to later times with increasing depth, as is indicated by Equation 2.8. (Courtesy of C. Tongyai)

ferentiating Equation 2.8 with respect to z, setting $z = 0$, and substituting this into Equation 2.3 to obtain

$$G(0) = \rho_s c_s A(0) \sqrt{\kappa\omega} \sin (\omega\tau + \pi/4). \tag{2.9}$$

Equation 2.9 tells us that the maximum soil heat flux occurs

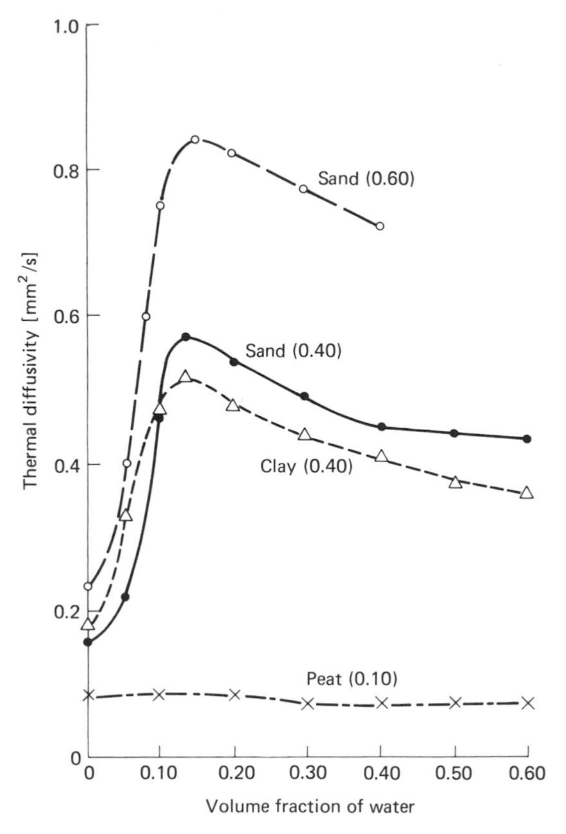

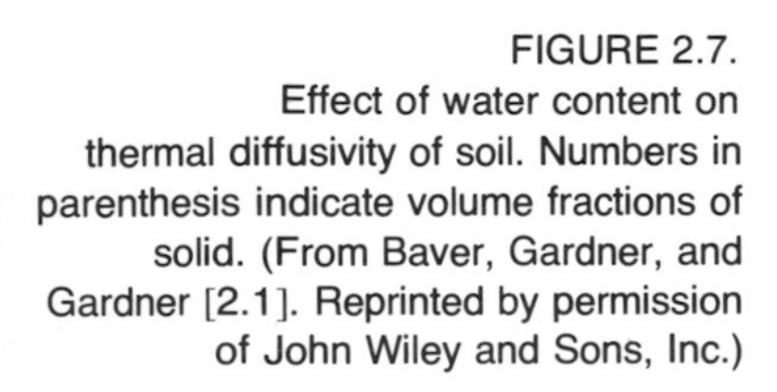
FIGURE 2.7. Effect of water content on thermal diffusivity of soil. Numbers in parenthesis indicate volume fractions of solid. (From Baver, Gardner, and Gardner [2.1]. Reprinted by permission of John Wiley and Sons, Inc.)

1/8 cycle before the maximum temperature. We already noted this at the beginning of the chapter, from experimental observations. If the maximum heat flux is at solar noon, the maximum temperature will occur three hours later.

Two points need to be made before we conclude this section on soil temperature. First, though the surface temperature can be approximated by Equation 2.7 for a clear day and constant κ with depth, fluctuations in heat input at the surface, especially on days with intermittent clouds, and changes in κ with depth cause rather large departures from the assumed sinusoidal wave. The damping depth concept, however, is still a useful one. If one knows the period of oscillation and the diffusivity, the depth of influence can be predicted.

The second point is that, though we have treated the surface temperature as an independent variable, it is actually determined by heat inputs and losses at the surface. A bare dry soil surface will therefore have large surface temperature fluctuations, while soil surfaces under vegetation will show much smaller fluctuations.

The theory we have developed here can be used to predict temperature ranges at different depths in the soil. A measurement of temperature at a single depth, along with an estimate of κ, can be used with Equation 2.8 to predict maximum and minimum temperatures at other depths. Since temperature is so important to the function of living systems, and the range of soil temperature varies so drastically with depth, it stands to reason that prediction of temperature as a function of depth in soil would be extremely important to studies of soil biology.

References

2.1 Baver, L. D., W. H. Gardner, and W. R. Gardner (1972) *Soil Physics,* 4th Ed. New York: Wiley.

2.2 Geiger, R. (1965) *The Climate Near the Ground.* Cambridge, Mass.: Harvard University Press.

2.3 Lawrence, R. S., G. R. Ochs, and S. F. Clifford (1970) Measurements of atmospheric turbulence relevant to optical propagation. *J. Opt. Soc. Am. 60*:826–830.

2.4 Lawrence, R. S., G. R. Ochs, and S. F. Clifford (1972) Use of scintillations to measure average wind across a light beam. *Appl. Optics 11*:239–243.

GENERAL REFERENCE FOR SOIL TEMPERATURE AND HEAT FLOW

Van Wijk, W. R. (ed.). (1963) *Physics of the Plant Environment.* New York: Wiley.

Problems

2.1 Using the following midday temperature data
a. Plot height as a function of mean temperature
b. Plot ln $(z + z_H - d)$ (Equation 2.2 as a function of mean temperature; assume $d = 0$ and $z_H = 0.01$ m.
c. From the plot in (b), find the surface temperature (temperature at $z = d$)

Height [m]	Mean air temperature [°C]
6.4	31.23
3.2	31.86
1.6	32.44
0.8	33.43
0.4	34.41
0.2	35.25
0.1	36.24

2.2 A common saying among farmers is "a wet soil is a cold soil." Is this true? At what water content would you expect a mineral soil to warm fastest (or have the largest damping depth)?

2.3 If the daily maximum and minimum soil temperatures at the soil surface are 35°C and 15°C, respectively, plot the maximum and minimum temperatures as a function of depth to 30 cm for a clay soil at 0.2 m^3/m^3 water content.

3 Environmental Moisture

Environmental moisture is of concern for two reasons. First, the biochemical reactions that sustain life in biological systems go on in water. Organisms are seldom in moisture equilibrium with their surroundings. Since maintenance of proper internal water balance is of great importance to the functioning and survival of organisms, we need to understand the laws governing rates of moisture transfer between organisms and their surroundings.

Second, environmental moisture is important in energy transport. If there is a change of phase associated with water transport, large quantities of energy can be transferred to or from a surface. Evaporation of 1 cm depth of water from 1 cm^2 of surface would require about 2.5 kJ, approximately the energy delivered to a square centimeter of surface by the sun during an entire clear summer day. If water vapor were to condense on a surface and freeze, about 2.8 kJ of heat per gram of water would be liberated through condensation and freezing. The addition of the latent heat of condensation to air, once the air is cooled to the dew point temperature, can sometimes prevent air temperature from dropping much below the dew point temperature. Orchard frost warnings are often based on this principle.

To find either the rate of latent heat loss or the rate of water loss it is necessary to be able to calculate water vapor flux. This is given by Fick's law, which, in integrated form is

$$E = \frac{\rho_{vs} - \rho_{va}}{r_v} \tag{3.1}$$

where E is the water vapor flux density (g m^{-2} s^{-1}), r_v is the resistance to vapor diffusion (s/m) and ρ_v is the vapor density (g/m^3). The vapor diffusion resistance (r_v) depends on the properties of the exchange surface and the air flow. The ''driving force'' for evaporation is the difference between vapor density at the evaporating surface (ρ_{vs}) and in the surrounding air (ρ_{va}). To find this difference, we need to know how to measure or estimate vapor density.

Saturation Conditions

If a container of pure water is uncovered in a closed space, water will evaporate into the space from the liquid water. As water evaporates, the concentration of water molecules in the gas phase will increase. Finally an equilibrium will be established when the number of molecules escaping from the liquid water equals the number being recaptured by the liquid. If the liquid temperature were increased, the random kinetic energy of the molecules would increase, and more water would escape. At equilibrium, the vapor density or concentration would be higher over the warmer water. This equilibrium vapor density, established between liquid water and water vapor in a closed system, is known as the *saturation vapor density* for the particular temperature of the system. It is the highest concentration of water vapor that can exist in equilibrium with a plane, free water surface at a given temperature. The saturation vapor density is shown as a function of temperature in Figure 3.1. This relationship is independent of any other gases present in the space above the water. Tables giving saturation vapor density as a function of temperature can be found in List [3.2] and in Appendix A.3.

Water vapor content of air can also be expressed in terms of the partial pressure exerted by the water vapor, or *vapor pres-*

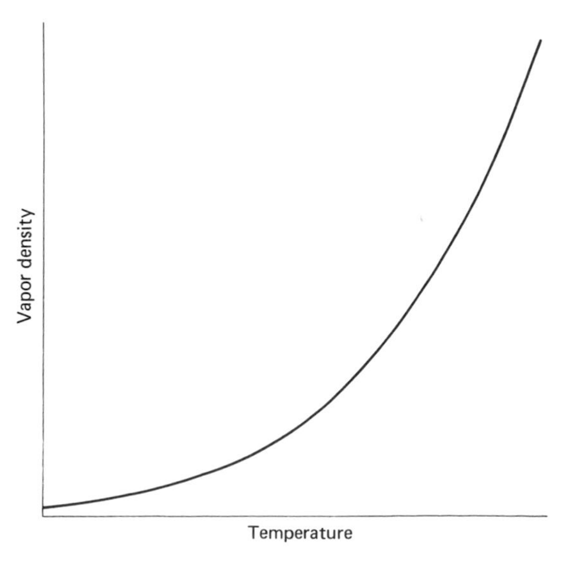

FIGURE 3.1. Density of water vapor in equilibrium with liquid water as a function of temperature.

sure. Vapor pressure is related to vapor density by the perfect gas law. For water vapor it is

$$p = 4.62 \times 10^{-4} \rho_v T.$$

Vapor pressure is in kilopascals when ρ_v is in g/m^3 and T is in Kelvins. Tables of p as a function of temperature are available (List [3.2]), but with modern scientific pocket calculators it is sometimes more convenient to use an equation. One which has sufficient accuracy for most purposes is

$$p = \exp\left(52.57633 - \frac{6790.4985}{T} - 5.02808 \ln T\right).$$

Other, more accurate equations are given by Riegel, *et al.* [3.3].

Conditions of Partial Saturation

In nature air is seldom saturated, so we need some means of expressing partial saturation. We can do this by giving the actual or ambient vapor density, or by specifying *relative humidity,* which is the ratio of ambient vapor density to the saturation vapor density at air temperature:

$$h_r = \frac{\rho_v}{\rho_v'} . \tag{3.2}$$

Relative humidity is sometimes multiplied by 100 to express it as a percent rather than a fraction. The relationship between saturation vapor density and ambient vapor density at various humidities is shown in Figure 3.2. For example, if air temperature were 27°C and the relative humidity 0.4, Figure 3.2 shows that the vapor density would be 10.4 g/m^3 (the intersection of $T_a = 27, h_r = 0.4$).

The *dew point temperature* is the temperature at which air, when cooled without changing its water content, just saturates. This can also be determined from Figure 3.2. For example, if air at 27°C and $0.4 h_r$ were cooled at constant ρ_v, the temperature at which the air would saturate (intersection of $h_r = 1$ and $\rho_v = 10.4$ g/m^3 lines) would be 11.4°C. This would be the dew point temperature, T_d. We see that any combination of two of the three variables T_a, T_d or ρ_v, and h_r uniquely determines the others.

One other moisture variable is of interest to us in describing atmospheric moisture; that is the *"wet bulb" temperature,* T_w. We are interested in T_w not only because it provides a means for measuring atmospheric moisture using a sling psychrometer, but also because the wet surface of an organism can act like a wet bulb heat exchanger under some conditions.

To understand the meaning of T_w, we might ask ourselves how we could determine the temperature drop which could be

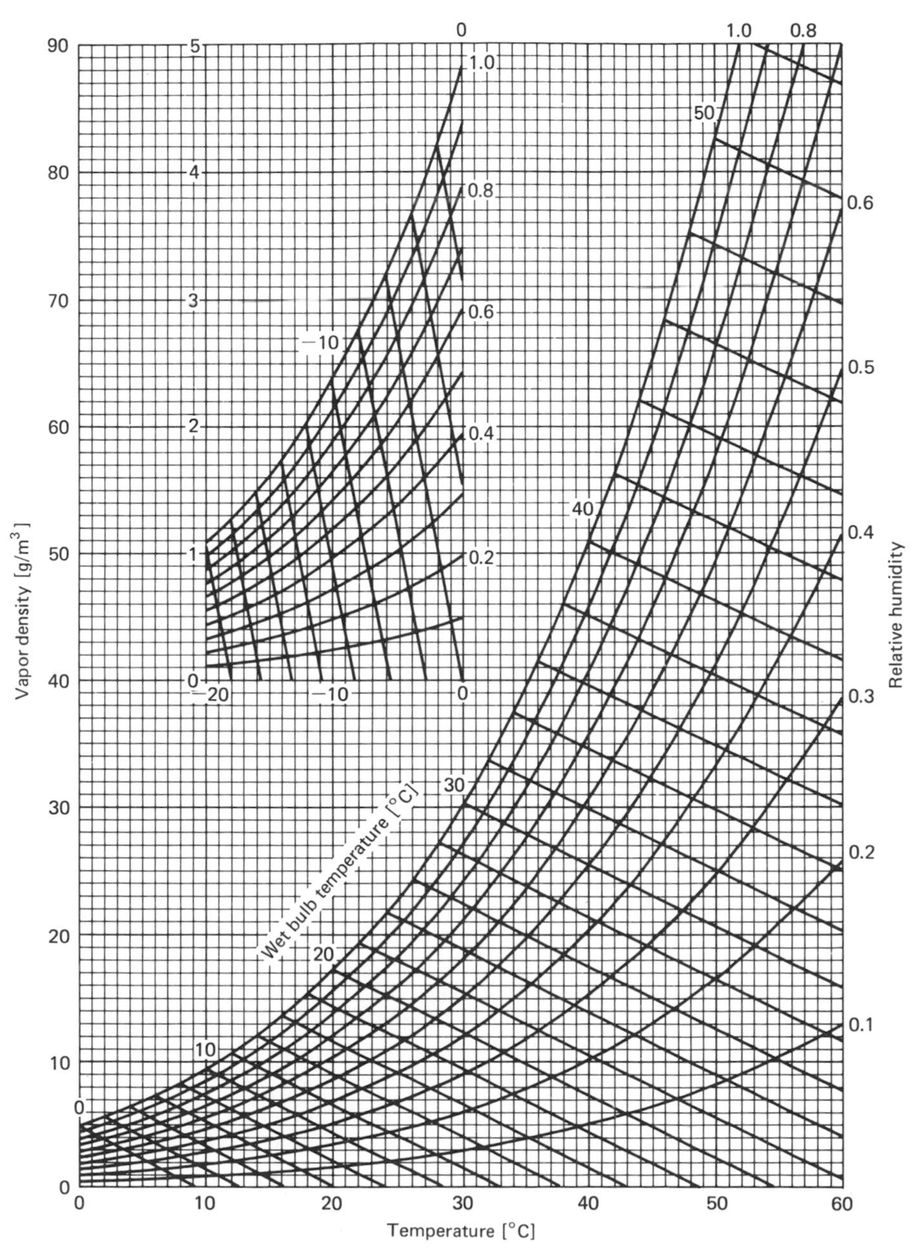

FIGURE 3.2. Temperature-vapor density-relative humidity diagram for sea level atmospheric pressure. The inset is for temperatures below 0 °C. Diagonal lines are for wet bulb temperature, and are spaced at 2 °C increments.

achieved by adiabatic evaporation of water into air. Air is cooled by evaporating water into it, but the evaporation of water into the air raises its vapor density. Since the change in heat content of the air due to changing its temperature must equal the latent heat of evaporation for the water evaporated into the air, we can write

$$\rho c_p(T_a - T_w) = \lambda[\rho'_{vw} - \rho_v] \tag{3.3}$$

where λ is the latent heat of vaporization for water, ρ is the density of the air, c_p is its specific heat, and ρ'_{vw} is the saturation vapor density at T_w.

Equation 3.3 can be rearranged to give the psychrometer equation:

$$\rho_v = \rho'_{vw} - \gamma(T_a - T_w) \tag{3.4}$$

with γ (= $\rho c_p/\lambda$) defined as a thermodynamic psychrometric "constant." The value for γ depends on temperature and atmospheric pressure. At 20°C and 100 kPa (sea level), λ = 0.495 g $m^{-3}K^{-1}$. Equation 3.4 defines the family of straight, diagonal lines shown in Figure 3.2. As an example, we could enter Figure 3.2 at T = 27°C and h_r = 0.4. Following the diagonal line to the wet bulb temperature scale we find that T_w = 17.9°C and ρ'_{vw} = 15.3 g/m³. An ideal evaporative cooler or sling psychrometer would therefore cool to 17.9°C under those conditions. Cooling of actual psychrometers and coolers may be less because cooling is not strictly adiabatic in these devices, but the ideal is approached with a well ventilated sling psychrometer.

Vapor Densities in Nature

Generally evaporation and transpiration rates are higher in the day than at night so vapor densities are highest near the ground and during the day, but often variations are less than 1–2 g/m³. Typical diurnal and spatial behavior of vapor density on summer days is shown in Figure 3.3. During the day, the vapor density is high near the surface, and decreases with height. At night vapor density may increase with height, indicating dewfall. Within several meters of the ground surface, vapor density is a logarithmic function of height. An equation similar to Equation 2.2 applies under conditions that will be discussed later, so a log plot of two or more measured vapor densities with height would allow extrapolation or interpolation to other heights, as was done with temperature. However, in practice, this is seldom necessary because the changes in vapor density with height are usually relatively small. Actual measurements of vapor concentrations in air as functions of height and time of day can be found in Geiger [3.1].

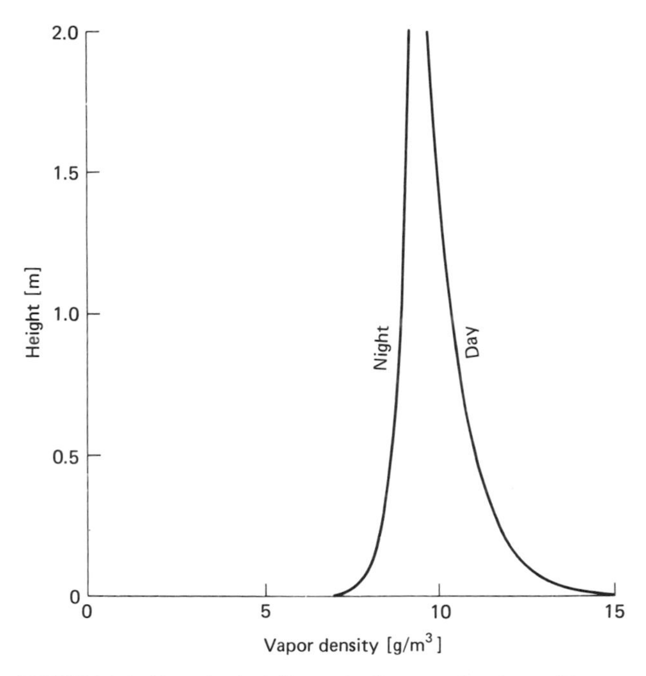

FIGURE 3.3. Hypothetical (but typical) vapor density profiles above a wet soil or crop surface on clear days and nights. The nighttime profile represents dewfall.

In the absence of airmass changes the vapor density often remains relatively constant over long periods of time (days to weeks) because it reflects the absolute amount of water in the air. Relative humidity, however, changes drastically from day to night, over time, and with height, since it depends on both vapor density and temperature (saturation vapor density). Since water vapor flux depends on vapor density gradients and not on humidity, it is generally best to measure atmospheric water content in terms of vapor density.

Very near evaporating surfaces the humidity may approach 1.0. Anyone who has walked through a tall crop, such as corn, on a hot day can verify this. High humidity is particularly important to growth of some disease organisms and survival of some insect eggs and larvae. Control of diseases and insect attacks on plants and animals can sometimes be effected by reducing the humidity in the microenvironment to levels not tolerated by the invading pest.

Liquid-phase Water

Water movement in soil and in organisms is mostly in the liquid phase. The driving forces and resistances to flow in the liquid phase are quite different from those in the vapor phase. To understand liquid-phase water exchange and organism response, it is necessary to have good physical descriptions of these driving forces and resistances.

In its simplest form, the equation describing one-dimensional liquid flux can be written as

$$J_w = -k \frac{d\Psi}{dx} \tag{3.5}$$

where k is the hydraulic conductivity ($kg^2m^{-1}s^{-1}J^{-1}$), J_w is the flux density ($kg\ m^{-2}s^{-1}$), and Ψ is the water potential (J/kg). The water potential is defined as the potential energy per unit mass of water with reference to pure water at zero potential. In thermodynamic terms it is the partial specific Gibbs free energy of the water in the system. The gradient of the water potential, $d\Psi/dx$ is the driving force for water movement in the liquid phase.

The water potential is made up of several components, some of which may not be active in a system at a particular time. Whether or not a component is active will depend on the system, as we will see. The components of water potential can be listed as

$$\Psi = \Psi_g + \Psi_m + \Psi_p + \Psi_o \tag{3.6}$$

where Ψ_g (gravitational) is the component due to position in the gravitational field, Ψ_o (osmotic) is the component due to dissolved solutes in the presence of a semipermeable membrane, Ψ_m (matric) is the component due to the attraction of the matrix for water molecules, and Ψ_p (pressure) is the component due to hydrostatic or pneumatic pressure.

The gravitational potential of free water at some level, z, above or below a reference level (up is +) is just

$$\Psi_g = gz \tag{3.7}$$

where g is the gravitational acceleration (9.8 m/s²). You will recognize this as the equation for potential energy (per unit mass) in a gravitational field.

The osmotic potential is equal to the potential energy difference (per unit mass of water) between water in solution and pure, free water at the same elevation. The osmotic component is only a driving force for water movement when solute movement is restrained with a semipermeable membrane, since no potential energy difference could be generated without restraining the solutes. The osmotic potential is approximated by the van't Hoff relation

$$\Psi_o \simeq -CRT \tag{3.8}$$

where C is the solution concentration in moles of ions (for electrolytes) or molecules (for nonelectrolytes) per kilogram of water, R is the gas constant (8.31 J $K^{-1}mol^{-1}$), and T is the Kelvin temperature. Note that Ψ_o can only be zero or negative.

The matric component arises from the attraction between water and soil particles, proteins, cellulose, etc. Adhesive and cohesive forces bind the water and reduce its potential energy (per unit mass) compared with pure, free water at the same elevation. For any substance that imbibes water there exists a relationship between water content and matric potential. This relationship is called a characteristic curve. Typical soil moisture characteristics are shown in Figure 3.4. Note that Ψ_m is always negative. Similar curves could be drawn for cellulose or protein.

The pressure potential in a soil or plant system is

$$\Psi_p = \frac{P}{\rho_w}$$

where P is the hydrostatic or pneumatic pressure in the system and ρ_w is the density of water. The pressure potential can be either positive or negative. It is important in plant cells, in the bloodstreams of animals, and in saturated soil. In a plant cell, the total water potential is the sum of the osmotic and pressure

FIGURE 3.4. Typical soil moisture characteristic curves for three soil textures.

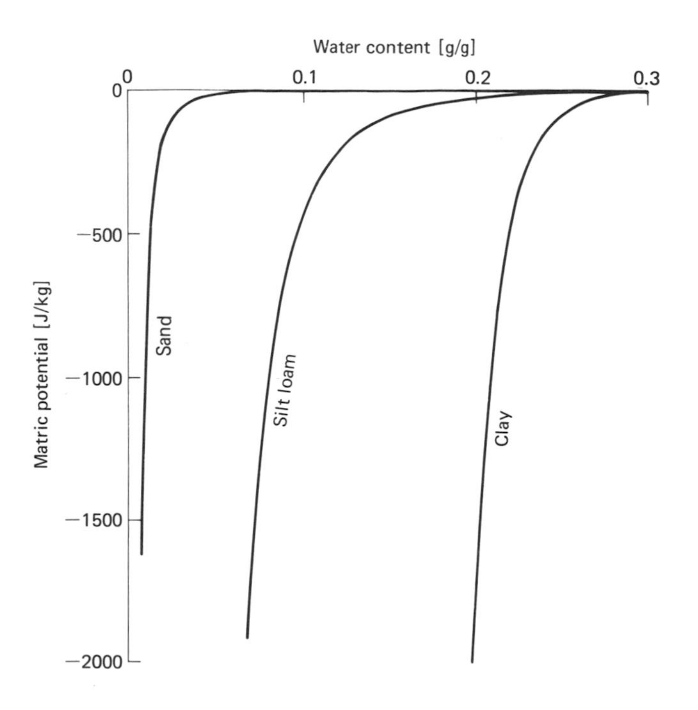

potentials. In a blood capillary, solutes are free to diffuse through the walls and tissues but proteins are not. The positive hydrostatic pressure in the blood is just balanced by the negative matric component due to proteins. If it were not for the matric component of blood protein, blood would move out into the tissues.

Relation of Liquid- to Gas-phase Water

We have talked about driving forces on liquid and on gas-phase water. There are a number of interfaces where these meet, and we need some means of determining the vapor density at evaporating surfaces of organisms from the water potential in the liquid phase. Once we know the vapor density at the evaporating surface we can use Equation 3.1 to determine the rate of evaporation from an organism. If we know the temperature of the evaporating surface, we can determine its saturation vapor density from Figure 3.2. If, in addition, we know the humidity at the surface, we can use Equation 3.2 to find ρ_{vs}. The relative humidity at the evaporating surface can be related to water potential of the liquid phase at the surface by considering the energy required to create a volume, dV, of water vapor. The first law of thermodynamics for an adiabatic system says:

$$d\,(\text{Energy}) = -p\,dV.$$

From the ideal gas law we can write

$$pV = nRT$$

and

$$dV = -\frac{nRT}{p^2}\,dp$$

so

$$d\,(\text{Energy}) = \frac{nRT}{p}\,dp.$$

The change in energy in going from a reference state where $p = p_0$, the saturation vapor pressure, to some other pressure, p is

$$\text{Energy} = nRT\int_{p_0}^{p} \frac{dp}{p} = nRT \ln \frac{p}{p_0}$$

and

$$\Psi = \frac{\text{energy}}{\text{mass}} = \frac{RT}{M} \ln\left(\frac{p}{p_0}\right)$$

where M is the molecular weight (0.018 kg/mole) of water.

RT/M for water at 25°C is 1.37×10^5 J/kg, so

$$\Psi = 1.37 \times 10^5 \ln \frac{p}{p_0}. \tag{3.9}$$

Within the range of water potentials commonly encountered in biological systems, ln p/p_0 can be approximated by $p/p_0 - 1$, and Equation 3.9 becomes

$$\Psi \simeq 1.37 \times 10^5 \left(\frac{p}{p_0} - 1\right). \tag{3.10}$$

Note that $p/p_0 = h_r$, so now we have the desired relationship between relative humidity at the evaporating surface and water potential of the liquid phase. To give some idea of relative humidities at evaporating surfaces, we will use Equation 3.10 to calculate h_r for some typical water potentials.

Typical water potentials of soils and plants are as follows. If a uniform, deep soil were wetted and allowed to drain freely for several days, we might expect to find Ψ_m around -30 J/kg. If plants were grown on the soil until they wilted and died from water stress, Ψ_m would be -2000 to -4000 J/kg, depending on plant species. The plant leaves, under normal growing conditions might have $\Psi_{\text{leaf}} = -1000$ to -2000 J/kg during the day and -200 to -300 J/kg during the night. As the plant stresses, Ψ_{leaf} becomes about equal to Ψ_{soil}. The relative humidity at -4000 J/kg would be (Equation 3.10)

$$h_r = 1 + \frac{\Psi}{1.37 \times 10^5} = 0.97.$$

Water potentials of animal body fluids are even higher (less negative) than this, so h_r at animal evaporating surfaces is greater than 0.97. Thus, for purposes of calculating vapor-phase transport, the relative humidity of evaporating surfaces of moist soils, leaves, and animals is almost always close enough to one to assume that they are at saturation vapor density. There are two notable exceptions. Some fungi and insects can survive at water potentials as low as -50 to -70 kJ/kg. This corresponds to relative humidities of 0.6 to 0.7. Thus "air dry" soil (soil in moisture equilibrium with air) in some locations would have water potentials high enough to support some biological activity.

The other exception that is of interest to us here is evaporation of sweat from skin. If sweat evaporation is sufficiently rapid that loss of liquid sweat from the skin is prevented, the solute (mainly NaCl) concentration at the skin will increase until evaporation is taking place from a saturated NaCl solution. The relative humidity of such a solution is 0.75, so $\rho_{vs} = 0.75\ \rho'_{vs}$. Assuming a skin surface temperature of 36°C, air vapor density of 25 g/m^3, and r_v of 100 s/m, the evaporation rate would be 0.16 g m^{-2}s^{-1} for clean skin and 0.06 g m^{-2}s^{-1} for salt-covered skin. The presence of salt would thus reduce evaporative cooling to less than half the potential rate. Overproduction of sweat, which results in sweat dripping from the skin, is considered by some to be wasteful, but nature is seldom wasteful when all factors are considered. We see from this example that the removal of salt from the skin surface by excess sweat production can be extremely important in increasing latent heat loss from animals.

References

3.1 Geiger, R. (1965) *The Climate Near the Ground*. Cambridge, Mass.: Harvard University Press.

3.2 List, R. J. (1971) *Smithsonian Meterological Tables*, 6th ed. Washington, D.C.: Smithsonian Institution Press.

3.3 Riegel, C. A., A. N. Hull, and T. M. L. Wigley (1974) Comments on "a simple but accurate formula for the saturation vapor pressure over liquid water." *J. Appl. Meteor. 13*:606–608.

3.4 Taylor, S. A., and G. L. Ashcroft (1972) *Physical Edaphology: The Physics of Irrigated and Nonirrigated Soils*. San Francisco: W. H. Freeman.

GENERAL REFERENCE ON WATER POTENTIAL

See Chapter 7 of Reference [3.4].

Problems

3.1 Use Figure 3.2 to fill in the following table

T_a	T_d	T_w	ρ_v	h_r
30	10			
22		14		
8			2	
−10				0.8

3.2 What is the rate of water loss from your skin surface just after you get out of the swimming pool? Assume r_v = 100 s/m, T_{skin} = 33°C, T_a= 30°C, and h_r= 0.2. Also find the rate of latent heat loss, taking the latent heat of vaporization as 2.43 kJ/g.

3.3 What is the water potential or component in each of the following?

a. Gravitational potential at the base of a 50-m tree with z = 0 at the top of the tree

b. The pressure potential in cells of a leaf if $\Psi_m = \Psi_g = 0$, $\Psi = -2300$ J/kg, $\Psi_0 = -3000$ J/kg

c. The osmotic potential of sea water if it is equivalent to 0.5 molal NaCl

d. The matric potential of a silt loam soil at a water content of 0.1 (Figure 3.4)

3.4 Food spoilage microorganisms cease growth at humidities around $h_r = 0.65$. What is their water potential? (See Equation 3.9.)

4 Wind

As living organisms, we are most acutely aware of three things about the wind. We know that it exerts a force on us and other objects against which it blows, it is effective in transporting heat from us, and it is highly variable in space and time. A fourth property of the wind, less obvious to the casual observer, but essential to terrestrial life as we know it, is its effective mixing of the atmospheric boundary layer of the Earth. This can be illustrated by a simple example. On a summer day about 10 kilograms of water are evaporated into the atmosphere from each square meter of vegetated ground surface. This would increase the vapor density in an air layer 100 meters thick by 100 g/m^3 if there were no transport out of this layer or condensation within it. The actual increase in vapor density in the first 100 meters of the atmosphere is only a few grams per cubic meter, so most of the water vapor has been transported to higher levels where clouds are formed. Monteith [4.7] does a similar calculation to show that photosynthesis in a normally growing crop would use all of the CO_2 in a 30-m layer above a crop in a day, yet measured CO_2 concentrations have diurnal fluctuations of only 15 percent or less. Without the vertical turbulent transport of heat, water vapor, CO_2, oxygen, and other atmospheric constituents, the microenvironment we live in would be very inhospitable indeed.

In order to determine the force of the wind on, or the rate of heat transfer from living organisms in their microenvironments, it is necessary to know the wind speed in the vicinity of the organism. This requires an understanding of the behavior of average wind in the surface boundary layer of the Earth.

The behavior of the wind, in turn, is dictated by rates of turbulent transfer in the surface boundary layer. Turbulent transfer theory will allow us to derive equations for wind, temperature, vapor density, and CO_2 profiles and fluxes, and will be helpful later when we discuss plant canopies and their environment. In this chapter we will first discuss the behavior of the wind in natural, outdoor environments. We will then present some of the fundamentals of turbulent transport theory and derive the profile equations for wind, temperature, and vapor density. Finally we will discuss the effect of buoyancy on transport, and wind within crop canopies. Actual calculations of heat, mass, and momentum transport to living organisms will be left to Chapter 6.

Characteristics of Atmospheric Turbulence

As was previously mentioned, one of the obvious characteristics of wind is its variability. We are aware of random variations of wind in time through fluttering of flags and leaves, variations in the force of wind on us, and other common experiences. Spatial variations are obvious when one looks at a field of ''waving grain'' or at ''cat's paws'' on a lake. We are also aware that the range of variability is large. We see very small scale fluctuations in ''heat waves'' on hot summer days, and feel or hear the effects of very large scale fluctuations as wind gusts, which blow dust or shake the house. All of these characteristics of wind with which we are intimately acquainted are characteristics of turbulent flow. Except for a thin layer of air close to surfaces, the atmosphere is essentially always turbulent, or, in other words, characterized by random fluctuations in wind speed and direction caused by a swirling or eddy motion of the air. These swirls or eddies are generated in two ways. As wind moves over natural surfaces, the friction with the surface generates turbulence. This is called mechanical turbulence. Turbulence is also generated when air is heated at a surface and moves upward due to buoyancy. This is called thermal or convective turbulence. The size of the eddies produced by these two processes is different, as is shown in Figure 4.1. The fluctuations from mechanical turbulence tend to be smaller and more rapid than the thermal fluctuations. A striking demonstration of these types of turbulence can be seen by watching the plume from a smokestack on a hot day. The plume is called a ''looping plume'' because, in addition to the small scale mechanical turbulence that tears the plume apart and spreads it with distance, the thermal updrafts and downdrafts cause the entire plume to be transported upward or downward. The smoke release in Figure 4.2 shows this behavior.

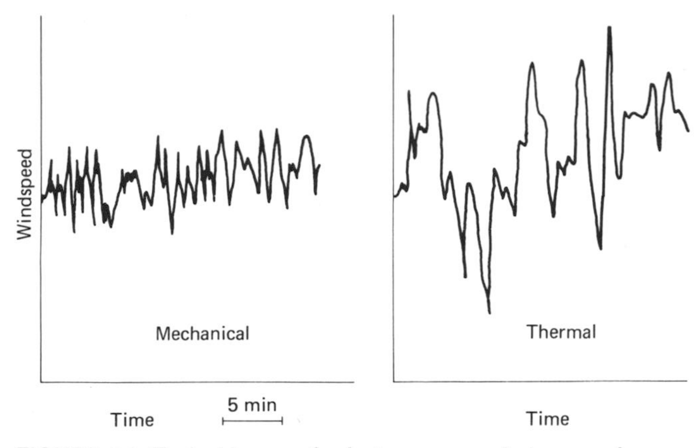

FIGURE 4.1. Typical traces of a fast-response wind sensor for conditions of pure mechanical turbulence, and mechanical plus thermal turbulence. (From Lowry [4.5].)

Large eddies, which are produced either mechanically or thermally, are unstable, and decay into smaller and smaller eddies until they are so small that viscous damping by molecular interactions within the eddies finally turns their energy into heat. The size of the smallest eddies produced by mechanical and convective motion (rather than breakdown of larger eddies) is called the "outer scale of turbulence." The eddy size at which significant molecular interaction (viscous dissipation) begins is called the "inner scale of turbulence." The outer scale is generally taken to be a few meters, and the inner scale, a few millimeters.

FIGURE 4.2. Photograph of a looping plume showing combined effects of mechanical and thermal turbulence. (From Gifford [4.3]; photograph courtesy of the Brookhaven National Laboratory.)

This process of larger eddies breaking down into ever smaller ones is expressed in a rhyme by L. F. Richardson, a scientist who is responsible for much of the theory of atmospheric turbulence (Gifford [4.3]):

Great whirls have little whirls
That feed on their velocity;
And little whirls have lesser whirls,
And so on to viscosity.

Gifford remarks that this parody of de Morgan's verse on fleas "may be the only statement of a fundamental physical principle in doggerel."

In order to talk intelligently about atmospheric processes, we need to separate out the random motion from the average drift, or wind. The wind velocity vector is commonly divided up into components along the axes of a rectangular coordinate system. For convenience, the coordinate system is oriented so that the x axis points in the direction of the mean wind. Wind velocity in the x direction is then taken as $u = \bar{u} + u'$, where $\bar{u}$ is the mean wind, averaged over a period ranging from 15 minutes to an hour, and u' represents fluctuations about this mean value. The lateral (side to side) velocity component (perpendicular to $\bar{u}$) is given the symbol v. Since the coordinate system is oriented so that $\bar{u}$ faces into the mean wind, $\bar{v}$, by definition, is zero, so $v = v'$. The vertical velocity is given the symbol w. Using an argument similar to that for v, we can also say that $w = w'$. Similarly, we can define the temperature or vapor density in terms of an average quantity and a fluctuating quantity. Examples of these fluctuations are shown in Figure 4.3.

The fluctuations, or eddies, in the atmosphere can be thought of, in a sense, as being like molecules in a gas. They bounce about with random motion, carried with the mean wind. It is these fluctuations that transport heat, water, momentum, etc. in the atmosphere. If a blob of air at one level, with a given temperature and momentum, jumps to a different level in the atmosphere, the old heat and momentum are carried to the new level. This is analogous to the diffusion process in a gas. It is, in fact, possible to measure the flux of heat, momentum, or mass by averaging the product of fluctuations of temperature, horizontal wind, or mass, and vertical wind. Thus

$$\tau = -\rho\overline{u'w'} \tag{4.1}$$
$$H = \rho c_p\overline{w'T'} \tag{4.2}$$
$$E = \overline{\rho_v' w'} \tag{4.3}$$

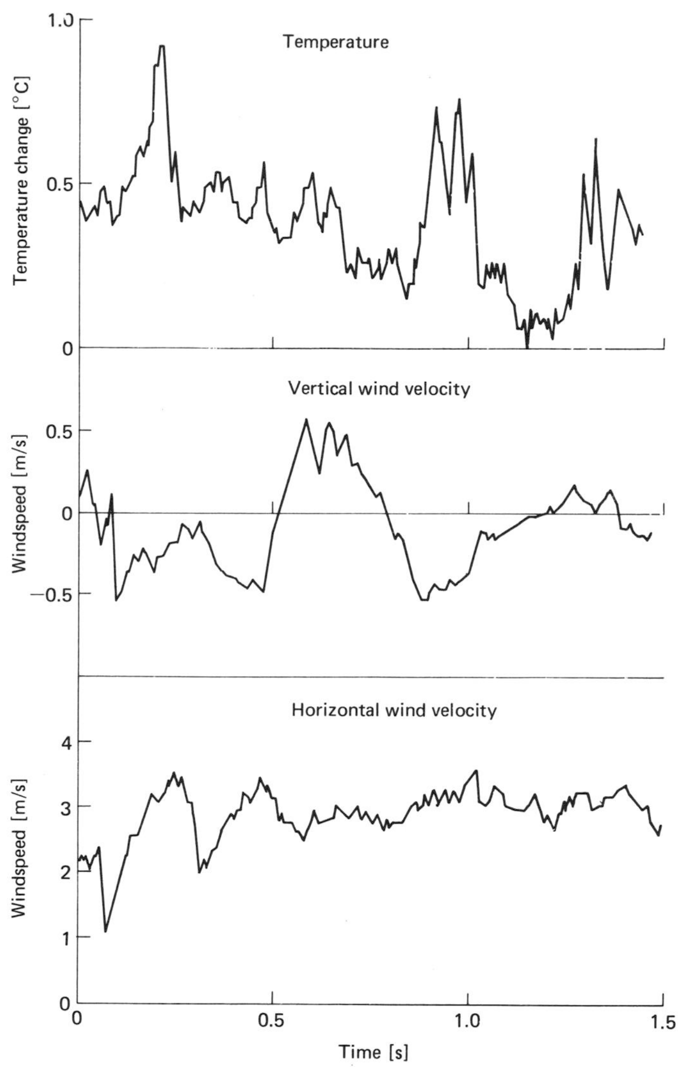

FIGURE 4.3. Fluctuations in horizontal and vertical wind and temperature recorded with fast-response sensors. (After Tatarski [4.8].)

where τ is the momentum flux to the surface, or drag of the wind on the surface, H is the heat flux, and E is the flux of water vapor. The overbars indicate averages taken over 15–30 minutes. Instrumentation problems make these equations somewhat difficult for routine measurements but they have been successfully applied in a number of studies [4.4].

In each of these flux equations, the transport is accomplished by fluctuations in the vertical wind component. The ability of the atmosphere to transport heat or mass depends directly on the size of the vertical fluctuations. Because of the proximity of the surface, the magnitude of w' is limited near the surface. Farther from the surface, transport is more efficient because the eddies are larger. We would therefore expect transport to increase with height. Increased intensity of turbulence will also increase vertical transport. Since turbulence is generated by mechanical action of wind moving over a rough surface, transport should increase as wind speed and surface roughness increase. We also know that turbulence is generated by buoyancy, so when there is strong heating at the surface, turbulent transport should increase. Strong cooling at the surface should result in reduced transport.

Flux and Profile Equations

We deal with turbulent transport in about the same way we would molecular transport problems. As was mentioned previously, the mechanism for transport is quite different, but we can use a similar mathematical approach. We simply define transport coefficients, K, such that

$$\tau = K_M \rho \frac{d\bar{u}}{dz} \tag{4.4}$$

$$H = -K_H \rho c_p \frac{d\bar{T}}{dz} \tag{4.5}$$

$$E = -K_v \frac{d\bar{\rho_v}}{dz} \tag{4.6}$$

where K_M is the "eddy viscosity," K_H is the "eddy thermal diffusivity," and K_v is the "eddy vapor diffusivity." These are steady-state flux equations for the surface boundary layer of the atmosphere. They are not very useful in this form because we have no way of knowing what values to give the K coefficients. From the discussion on turbulent mixing in the surface boundary layer we know that K will increase with height above the surface, wind speed, surface roughness, and heating at the surface. In the surface boundary layer, at steady state, we assume the flux densities, τ, H, and E, to be independent of height. Increases in the K coefficients with z will therefore be balanced by corresponding decreases in the gradients.

Let's put off discussing surface heating effects for the moment and concentrate on wind speed, roughness, and height effects on mixing. The wind speed and roughness effects are combined in the shear stress term, τ, but we will find it more convenient to define a new parameter called *friction velocity* as

u^* (m/s) $= (\tau/\rho)^{1/2}$. The meaning of u^* will become apparent soon.

If we assume that K is equal to some value, characterized by surface properties, at the exchange surface ($z = d$) and increases linearly with u^* and z, we can write

$$\begin{aligned} K_M &= ku^*(z + z_M - d) \\ K_H &= ku^*(z + z_H - d) \\ K_v &= ku^*(z + z_v - d) \end{aligned} \tag{4.7}$$

where the roughness parameters are functions of the exchange surface properties, and k is von Karman's constant, generally taken as equal to 0.4.

If Equation 4.7 for K_M is substituted into Equation 4.4 and the resulting equation integrated from the height of the exchange surface, d, to some height, $(z + z_M - d)$, the resulting equation describes $\bar{u}$ as a function of height. The equation is

$$\bar{u} = \frac{u^*}{k} \ln \frac{z + z_M - d}{z_M} \quad . \tag{4.8}$$

The other profile equations can be obtained similarly by integration as:

$$\bar{T} = T_o - \frac{H}{\rho c_p k u^*} \ln \frac{z + z_H - d}{z_H} \tag{4.9}$$

and

$$\bar{\rho_v} = \rho_{vo} - \frac{E}{ku^*} \ln \frac{z + z_v - d}{z_v} \tag{4.10}$$

where T_o and ρ_{vo} are the average values of surface temperature and vapor density at the exchange surface. Equations 4.9 and 4.10 describe the temperature and vapor profiles discussed in the previous two chapters.

The constant, d, is called the *zero plane displacement,* and can be thought of as the distance from the arbitrarily chosen height, zero, to the average height of heat, vapor, or momentum exchange. For a smooth surface, with z measured from the surface, $d = 0$. For dense vegetation (agricultural crops), d can be estimated from the average crop height, h:

$$d = 0.64h. \tag{4.11}$$

If the roughness elements are more sparsely spaced, Equation 4.11 does not hold. The d values probably differ for heat, momentum, and vapor, but the relationships have not been worked out. The *momentum roughness parameter,* z_M, is a length characteristic of the form drag at the momentum exchange surface. It depends on the shape, height, and spacing of the roughness elements. Again, empirical correlations provide us with an estimate of z_M for uniform surfaces:

$$z_M = 0.13h. \tag{4.12}$$

Other formulas that take into account the form and spacing of the elements must be used for more complex surfaces [4.1].

The roughness parameters for the other profile equations can generally be expressed as functions of the momentum roughness parameter. For our purposes we will use

$$z_H = z_v = 0.2\, z_M. \tag{4.13}$$

This relationship appears to work well for most vegetated surfaces, but should not be used for very smooth (ice, water, mud flat, etc.) surfaces. A more complete treatment of this is given by Garratt and Hicks [4.2].

As has been pointed out previously, these profile equations are useful for extrapolating or interpolating to find wind, temperature, vapor concentration, or CO_2 concentration at heights where they were not measured. This is shown for wind in Figure 4.4. The actual wind profiles form curved lines, when plotted in conventional linear fashion, but when $\ln(z + z_M - d)$ is plotted as a function of wind, a straight line results,

FIGURE 4.4. Wind profiles plotted as a function of height (left) and logarithm of height (right) for tall and short crops. The dashed lines on the log plot show the extrapolation of the measured profile to zero wind speed to determine the momentum roughness parameter.

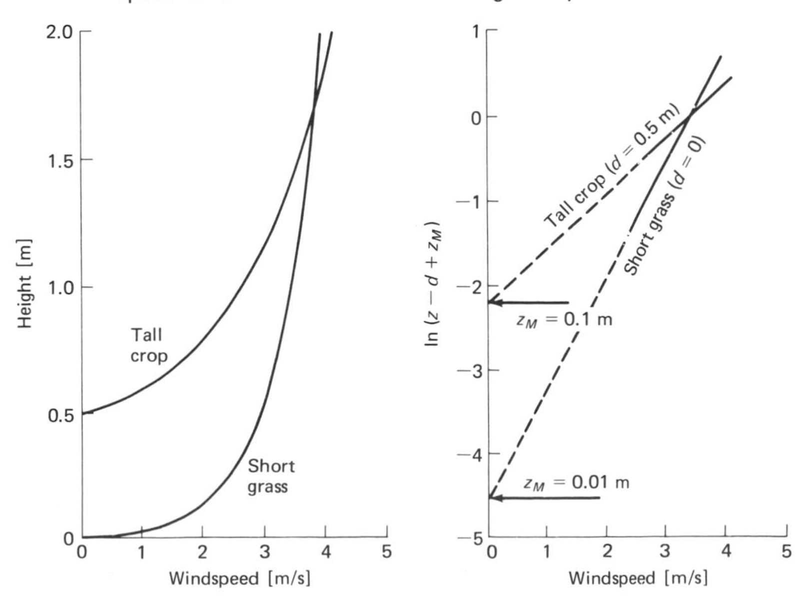

and measurements can be extrapolated to other heights. At least two measurements are required for the extrapolation, and if accuracy is desired, more are required. If the heights at which wind is measured are much greater than the value for z_M, ln $(z - d)$ can be plotted rather than ln $(z - d + z_M)$. Extrapolation of such plots to $u = 0$ gives an experimentally determined estimate of z_M. For rough estimates, it is possible to use Equations 4.11 and 4.12 to estimate z_M and d, and construct a complete wind profile from a single wind measurement. Though this is rough, it is often adequate for long-term estimates of environmental wind for plants and animals.

As an example, assume that the average wind at 2 m height is 3 m/s over a grass surface with average height of 20 cm. We would like to know the average wind speed in the vicinity of a sparrow on a fence wire 50 cm above the ground. Using Equations 4.11 and 4.12, we find that $d = 13$ cm and $z_M = 2.6$ cm. The ln term in Equation 4.8 equals 4.29, and $u^* = 0.28$ m/s. Using this value of u^* in Equation 4.8, we find that the wind at 50 cm is 1.9 m/s.

We should note in passing that Equations 4.9 and 4.10 can be used to find the heat or vapor flux in the boundary layer if temperatures or vapor densities at a minimum of two heights are measured, and if enough wind data are available to determine u^*. We will find these relationships useful later as we attempt to determine water loss from crops or forests.

Fetch and Buoyancy

Now that we have equations describing turbulent transport, we need to look briefly at the conditions under which we can expect them to apply. We started by assuming that the wind was "at steady state" with the surface (that there were no horizontal gradients). When wind passes from one type of surface to another it must travel some distance before a layer of air, solely influenced by the new surface, is built up. The height of influence increases with downwind distance. The length of uniform surface over which the wind has blown is termed *fetch,* and the wind can usually be assumed to be 90 percent or more equilibrated with the new surface to heights of $0.01 \times$ fetch. Thus, at a distance 1000 m downwind from the edge of a uniform field of grain, we might expect our wind profile equations to be valid to heights of around 10 m.

The effect of thermally produced turbulence on transport was alluded to earlier, but its quantitative description was not given. Our equations, so far, apply only when turbulence is produced mainly by mechanical means. Strong heating of the air near the Earth's surface causes overturning of the air layers, with resultant increases in turbulence and mixing. Conversely, strong cooling of these air layers suppresses mixing and turbulence. Thus convective production or suppression of turbulence is directly related to sensible heat flux (H) at the surface. When H is positive (surface warmer than air), the atmosphere is said to be unstable, and mixing is enhanced. When H is negative, the atmosphere is said to be stable, and mixing is suppressed by thermal stratification.

The main components of a (random) kinetic energy budget for a steady-state atmosphere can be written as [4.6]

$$\frac{-u^{*3}}{kz} + \frac{gH}{\rho c_p T} = \epsilon \tag{4.14}$$

where g is the gravitational acceleration. The first term represents mechanical production of turbulent kinetic energy, the second term is the convective production, and these two together equal the viscous dissipation of the energy. The ratio of convective to mechanical production of turbulence can be used as a measure of atmospheric stability [4.1]:

$$\zeta = \frac{-kzgH}{\rho c_p T u^{*3}} \tag{4.15}$$

The diabatic flux and profile equations can now be written as functions of stability and the other parameters we have discussed. Only the wind and temperature equations will be given. Fluxes of water vapor or CO_2 apparently obey equations similar to the heat flux equation, and profile equations for these variables are similar to the temperature profile equation. For the diabatic case, Equations 4.4 and 4.5 become

$$\tau = \frac{K_M}{\phi_M} \rho \frac{d\bar{u}}{dz} \tag{4.16}$$

and

$$H = -\frac{K_H}{\phi_H} \rho c_p \frac{d\bar{T}}{dz}. \tag{4.17}$$

The diabatic influence functions, ϕ_M and ϕ_H, are plotted as functions of atmospheric stability (ζ) in Figure 4.5. Note that as the atmosphere becomes unstable (ζ negative), ϕ decreases. This increases mixing and transport. A stable atmosphere has the opposite effect.

Diabatic profile equations corresponding to Equations 4.8 and 4.9 are

$$\bar{u} = \frac{u^*}{k}\left[\ln\left(\frac{z + z_M - d}{z_M}\right) + \psi_M\right] \tag{4.18}$$

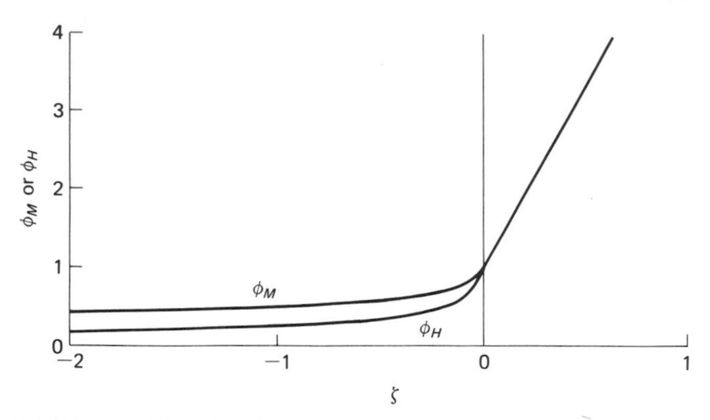

FIGURE 4.5. Diabatic influence functions for heat and momentum, plotted as functions of stability (ζ).

and

$$\overline{T} = T_o - \frac{H}{\rho c_p k u^*} [\ln\left(\frac{z + z_H - d}{z_H}\right) + \psi_H]. \tag{4.19}$$

The profile correction factors, ψ_M and ψ_H, are shown as functions of stability in Figure 4.6. Note that the correction is negative for unstable conditions, indicating that the wind at some height will be less than that predicted by a neutral profile equation. The converse is, of course, true for a stable atmosphere. This behavior seems quite reasonable since we expect greater coupling between layers of air in unstable flow and less coupling in stable flow than exists in neutral flow.

The stability corrections have little effect on the profiles one would predict for animal or plant environment studies, but are important when one wishes to measure fluxes in the surface boundary layer.

FIGURE 4.6. Diabatic profile correction factors for wind and temperature as a function of stability (ζ).

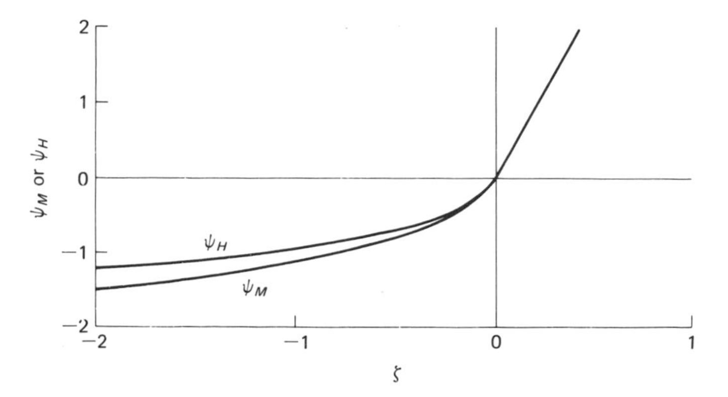

Wind within Crop Canopies

The equations given so far deal only with wind above the plant canopy. These equations are useful for determining the wind in the microenvironment of many living organisms, but cannot be used for leaves within a canopy or for animals that live within a crop or forest canopy. Prediction of wind within canopies is quite complicated, and we will only give an outline of the problem and a few useful formulas here. For more details, one can consult the paper by Bussinger [4.1].

The canopy flow regime can be thought of as being divided into three layers. The top layer, from the top of the canopy down to about the height of the zero plane ($z = d$), is the layer that exerts drag on the wind above the crop. The wind in this region is in the same direction as the mean wind, and the wind speed decreases exponentially with distance down from the top of the canopy in this layer. The second layer extends from the zero plane down to a height of 5 to 10 percent of the canopy height. In this layer, which is mainly in the stem space of the crop, the wind may be quite unrelated to the wind above the canopy, in both speed and direction. An example of the behavior of the wind in this intermediate layer can be observed by watching the drift of smoke from a campfire in a forest. This wind results from horizontal pressure differences within the canopy, and is attenuated by drag of the elements within

FIGURE 4.7. Wind profiles in crop and forest canopies, and profiles predicted by Equation 4.19 with various values of the attenuation coefficient, *a*. (After Bussinger [4.1].)

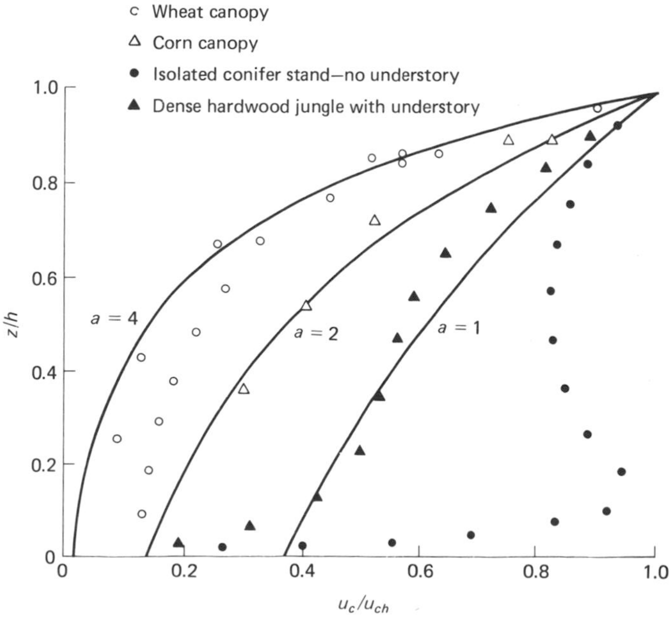

the stem space and by the ground surface. The third layer is a new logarithmic profile, similar in shape to the one above the canopy, with the wind speed at the top of this layer matching the wind at the bottom of layer two.

An equation that describes the wind speed within the top layer of the canopy, even in complex canopies, and works well for the top two layers in fairly uniform canopies is

$$u_c = u_{ch} \exp\left[a\left(\frac{z}{h} - 1\right)\right] \tag{4.20}$$

where u_{ch} is the windspeed at the top of the canopy, h is the canopy height, and a is an attenuation coefficient ranging from a number close to zero for very sparse canopies to around four for dense canopies. The wind at the top of the canopy can be estimated from the wind profile equations given previously. Figure 4.7 shows normalized wind profiles for both simple and complex canopies.

References

4.1 Bussinger, J. A. (1975) Aerodynamics of vegetated surfaces. *Heat and Mass Transfer in the Biosphere* (D. A. de Vries and N. H. Afgan, eds.) New York: Wiley.

4.2 Garratt, J. R. and B. B. Hicks (1973) Momentum, heat, and water vapor transfer to and from natural and artificial surfaces. *Quart. J. Roy. Meteor. Soc. 99*:680–687.

4.3 Gifford, F. A. (1968) An outline of theories of diffusion in the lower layers of the atmosphere. *Meteorology and Atomic Energy* (D. H. Slade, ed.). USAEC Division of Technical Information Extension, Oak Ridge, Tenn.

4.4 Hicks, B. B. (1971) The measurement of atmospheric fluxes near the surface: a generalized approach. *J. Appl. Meteor. 9*:386–388.

4.5 Lowry, W. P. (1969) *Weather and Life*. New York: Academic Press.

4.6 Lumley, J. L. and H. A. Panofsky (1964) *The Structure of Atmospheric Turbulence*. New York: Wiley.

4.7 Monteith, J. L. (1973) *Principles of Environmental Physics*. New York: American Elsevier.

4.8 Tatarski, V. I. (1961) *Wave Propagation in a Turbulent Medium*. New York: Dover.

Problems

4.1 If the wind, measured at 2 m height over a wheat field is 5 m/s, what is the wind speed at the top of the canopy? Assume the wheat is 60 cm high.

4.2 Using the following data, find z_M, z_H, u^*, H, and T_o. Assume d = 0.08 m and neutral stability.

Height [m]	Windspeed [m/s]	Temperature [°C]
0.5	7.0	26.0
1.0	8.3	25.4
2.0	9.4	24.8
4.0	10.6	24.2

4.3 Using the values from Problem 4.2 for u^* and H, find the stability parameter, ζ at $z = 2$m. Use Figures 4.5 and 4.6 to determine the stability corrections for fluxes and profiles. How much error in H results from assuming neutral conditions in Problem 4.2?

4.4 What value of a in Equation 4.20 would best fit the isolated conifer stand data in Figure 4.7?

5 Radiation

One of the most important modes of energy transfer between organisms and their environment is radiation. Radiant energy is transferred by photons, discrete bundles of energy that travel at the speed of light ($c = 3 \times 10^8$ m/s) and have properties similar to both particles and waves. These photons are emitted or absorbed by matter as a result of discrete quantum jumps in electronic energy levels in atoms or changes in vibrational and rotational energy levels in molecules. The wavelength of the radiation is uniquely related to the photon energy in an equation due to Planck: $e = hc/\lambda$, where h is Planck's constant (6.63×10^{-34} J s) and λ is the wavelength. Thus green photons, having a wavelength of 0.55 μm would have an energy $e = 6.63 \times 10^{-34}$ J s $\times 3 \times 10^8$ m/s $\div 5.5 \times 10^{-7}$ m $= 3.6 \times 10^{-19}$ J. This number is inconveniently small, and the main purpose of this type of calculation is to find the energy available for photochemical reactions anyway, so the energy contained in a mole (6.02×10^{23}) of photons is often calculated. A mole of photons is called an Einstein (E). The energy of photons at 0.55 μm is $6.02 \times 10^{23} \times 3.6 \times 10^{-19} = 2.2 \times 10^5$ J/E. The photon flux from a source can be found by dividing the energy flux at each wavelength of the source by the photon energy at the wavelength and summing over all wavelengths. If the spectrum of a source is reasonably continuous over a given waveband, the photon flux in that waveband can be calculated from the average energy over the waveband divided by the photon energy at the median wavelength. For example, photosynthetically active radiation (PAR) is generally considered to be radiation between 0.4 and 0.7 μm. For solar radiation, the median wavelength in the

0.4–0.7 μm waveband is 0.51 μm, for which the photon energy is 2.35×10^5 J/E. Near midday on a clear day the average energy flux density in this waveband is around 500 W/m^2, so the photon flux would be around 2.1×10^{-3} E $m^{-2}s^{-1}$.

We could also express the energy of photons as a function of frequency (v) of the radiation, since $v\lambda = c$ to give $e = hv$. Frequency rather than wavelength is used in some treatments of environmental radiation [5.4]. Advantages in using frequency are a more symmetrical presentation of absorption bands and ability to show both short- and long-wave radiation on a single graph. These advantages are offset somewhat by the loss of detail in the long-wave portion of the spectrum and the unfamiliar nature of the units to most biologists. In this presentation we will continue to use wavelength.

Blackbody Radiation

As was mentioned, photons are emitted or absorbed because of discrete energy transitions in the emitting or absorbing medium. Each allowable transition produces photons at a single wavelength. If there are many transitions at closely spaced energy levels, the lines tend to merge into an emission or absorption band. If there are an infinite number of transitions spaced throughout the electromagnetic spectrum, the medium is a perfect radiator or absorber. It will absorb all radiation falling upon it and will radiate the maximum amount of energy that a medium at its temperature is capable of radiating. Such a medium is called a "blackbody." No such material exists in nature, but some materials approach this behavior over parts of the electromagnetic spectrum. Thus we may speak of a blackbody radiator at visible wavelengths or a blackbody radiator at infrared wavelengths, but would not necessarily expect the same material to be a blackbody in both wavebands. Snow is a very poor absorber of visible radiation, but almost a perfect blackbody in the far infrared.

Definitions

In order for us to speak precisely and unambiguously about radiation, it is necessary for us to agree on definitions of some terms. Terms that we will use are listed below. Additional detail is available from Monteith [5.9].

Absorptivity (a): The fraction of incident radiation at a given wavelength that is sborbed by a material

Emissivity (ϵ): The fraction of blackbody emission at a given wavelength emitted by a surface

Reflectivity (r): The fraction of incident radiation at a given wavelength reflected by a surface

Transmissivity (t): The fraction of incident radiation at a given wavelength transmitted by a material
Radiant flux: The amount of radiant energy emitted, transmitted, or received per unit time
Radiant flux density (Φ): Radiant flux per unit area
Irradiance: Radiant flux density incident on a surface
Radiant emittance: The radiant flux density emitted by a surface

The relationships among these terms may be made somewhat clearer as follows: Assume a 2 m^2 source has a radiant emittance of 1 W/m^2. The radiant flux density at the source is 1 W/m^2 and the radiant flux of the source is 2 W. This radiation is emitted in all directions through a hemisphere surrounding the surface.

Once radiant energy arrives at a receiver it is either absorbed, transmitted, or reflected. Since all of the energy must be partitioned between these, we can write that $a + r + t = 1$. For a blackbody, $a = 1$, so $r = t = 0$.

We have seen that emission and absorption of radiation are linked by the same process—that of changing the energy status of the emitting or absorbing atoms or molecules. Thus we would expect the emissivity and absorptivity of a material at a given wavelength to be equal ($\epsilon = a$), which is a statement of Kirchhoff's law. It is important to recognize, at this point, that the absorptivity or emissivity values represent only fractions of possible absorption or emission at a particular wavelength, and say nothing about whether or not radiation is actually being absorbed or emitted at that wavelength. For example, carbon black has an emissivity and absorptivity for visible radiation of nearly one. When it is at room temperature, however, it emits negligible quantities of short-wave radiation, though it may be absorbing short-wave radiation from its surroundings. The radiant emittance is near zero, not because the emmisivity is low but because there is no energy to be emitted at visible wavelengths from such a cold surface.

Directional Relations

The irradiance at a surface depends on the orientation of the surface to the radiant beam. This is easily seen by considering the area on a surface covered by a beam of parallel light as the surface is inclined with respect to the beam. The radiant flux of the beam remains constant, but the beam covers a larger and larger area as the surface is inclined, so the flux density at the surface decreases. This is described quantitatively by Lambert's cosine law:

$$\Phi = \Phi_0 \cos \theta \tag{5.1}$$

where Φ_0 is the flux density normal to the beam, Φ is the flux density at the surface, and θ is the angle between the light beam and a normal to the surface.

The only common source of parallel light in natural environments is the sun. Lambert's law can be used to calculate the direct solar irradiance on a slope, a wall, a leaf, or an animal if we know Φ_0 and the angle the sun makes with a normal to the surface.

Attenuation of Radiation

Parallel monochromatic radiation propagating through a homogeneous medium that attenuates the beam will show a decrease in flux density described by Bouguer's law:

$$\Phi = \Phi_0 \, e^{-kx} \tag{5.2}$$

where Φ_0 is the unattenuated flux density, x is the distance the beam travels in the medium, and k is an extinction coefficient (m^{-1}) for the medium. This law has been used to describe light penetration in the atmosphere [5.2] in crop canopies [5.9], in water [5.10], and in snow [5.10]. The law applies only for wavebands narrow enough that k remains relatively constant over the waveband.

Radiant Emittance

The radiant energy emitted by a unit area of surface of a blackbody radiator can be calculated from the Stephan–Boltzmann law

$$\Phi_B = \sigma T^4 \tag{5.3}$$

where Φ_B is the emitted flux density (W/m^2), T is the Kelvin temperature, and σ is the Stephan–Boltzmann constant (5.67×10^{-8} W $m^{-2}K^{-4}$). Values of Φ_B at various temperatures are given in Table A.3. The Earth can be considered as approximating a blackbody radiator emitting at 288 K. The average emittance of the Earth is therefore 5.67×10^{-8} W $m^{-2}K^{-4} \times (288\ K)^4 = 390\ W/m^2$. The sun emittance can be approximated by that of a blackbody at 6000 K. The energy emitted is therefore 73 MW/m^2 at the sun's surface.

The energy emitted by nonblackbodies or "gray" bodies is given by

$$\Phi = \epsilon\sigma \, T^4 \tag{5.4}$$

where ϵ is the emissivity of the surface. For a blackbody, $\epsilon = 1$. Most natural surfaces have long-wave emissivities between 0.90 and 0.98. The emissivity is a function of wavelength, though it can often be treated as constant and equal to some average value for fairly large wavebands.

Spectral Distribution of Blackbody Radiation

Besides being concerned with the total energy radiated by a source, we need also be concerned with the spectral distribution of the radiant energy. Photochemical reactions in biological systems—such as photosynthesis, sight, and sunburn—respond only to radiant energy in limited wavebands.

The radiant energy spectrum from a blackbody is given by Planck's law

$$i_B = \frac{2\pi\, hc^2}{\lambda^5[\exp\,(hc/\lambda kT) - 1]}$$

where i_B (W/m³) is the energy flux density per unit wavelength, or spectral emittance, and k is the Boltzmann constant (1.38×10^{-23} J/K). Blackbody spectra are plotted in Figure 5.1 for sources at 6000 K and 288 K, corresponding roughly to sun and Earth emittance spectra. The total radiant emittance is the integral of Equation 5.5 over all wavelengths, and is given by Equation 5.3. The wavelength of peak spectral emittance is a function of temperature of the emitting surface, as can be seen from Figure 5.1. The relationship between wavelength of peak emittance (on a wavelength basis) and temperature is called the Wien law:

$$\lambda_m = \frac{2897}{T} \tag{5.5}$$

where λ is in μm and T in Kelvins.

Spectral Characteristics of Short-wave Radiation

Because the emitting gasses in the sun's atmosphere are at very high temperatures and pressures, the sun's spectrum is nearly continuous and approximates a 6000 K blackbody spectrum. The extraterrestrial solar spectrum is shown in Figure 5.2.

FIGURE 5.1. Blackbody spectral emittance for a 6000 K source (similar to the sun) and a 288 K source (similar to the Earth).

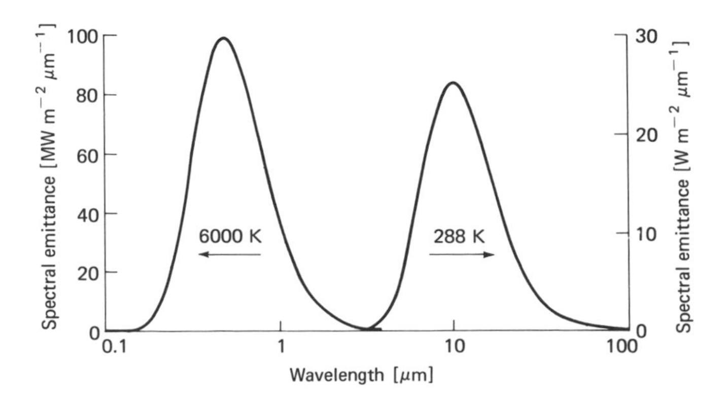

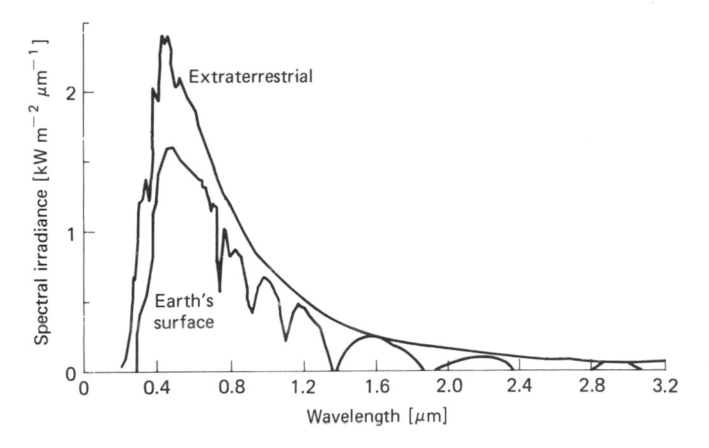

FIGURE 5.2. Spectral irradiance above and below the Earth's atmosphere. Atmospheric absorption in the ultraviolet is primarily by O_3, and in the infrared, primarily by H_2O. (After Gates [5.4].)

After the solar radiation passes through the Earth's atmosphere, radiation in some wavelengths has been almost completely absorbed. The ozone layer in the ionosphere absorbs much of the ultraviolet radiation, and water vapor and carbon dioxide have many strong absorption bands in the infrared (Figure 5.2).

Energy over the entire spectrum is reduced by Rayleigh (small-particle) and Mie (large-particle) scattering. Rayleigh scattering is most pronounced at short wavelengths, so the scattered radiation is blue. This is the source of the blue color of the sky. Mie scattering is generally most effective for longer wavelengths and results in the redness of scattered radiation from dust or haze.

About half of the solar energy arriving at the Earth's surface under a clear sky is in visible wavelengths (shorter than 0.7 μm) and about half in the infrared. Under cloudy skies the distribution of radiation between visible and infrared wavebands remains essentially the same as for clear skies, but radiation at both short and long wavelengths is strongly depleted.

The mean radiant flux density outside the Earth's atmosphere and normal to the solar beam is about 1.36 kW/m^2. This value fluctuates by $\pm$1.5 percent or so due to fluctuations in solar activity and $\pm$3.5 percent annually from changes in the Earth-to-sun distance.

Only a small percentage of the solar energy is at wavelengths longer than 4 μm and only a small part of the energy emitted by terrestrial sources is at wavelengths shorter than 4 μm (Figure 5.1). For convenience, we therefore divide environmental radiation into short-wave (0.3–4 μm) and long-wave (4–80 μm) portions in future discussions. The dividing line is, of course, arbitrary, but works well for our purposes. Thus, when we speak of short-wave radiation we will generally mean radiation (direct, reflected, or scattered) originating at the sun. Long-wave radiation will generally refer to radiation originating at terrestrial sources.

Soil, vegetation, water, etc. alter the spectral composition of radiation as it is differentially absorbed, reflected, or transmitted. This is obvious from the fact that we see colors characteristic of these surfaces (e.g., green trees and brown soil). Spectral absorptivities for a leaf and several animals are shown in Figure 5.3. The green color of leaves results from the relatively lower absorptivity in this portion of the spectrum than in the blue or red bands. The low absorptivity in the near infrared results in a reduction of the solar heat load on the leaf.

Radiant Fluxes in Natural Environments

Before going into a detailed discussion of radiant energy exchange between organisms and their surroundings, we need to consider the types of data required. Two types of energy exchange studies are of interest to ecologists. For the first, detailed observations of radiant flux densities to and from an organism are needed to compute a detailed energy budget. These detailed observations must be obtained by direct mea-

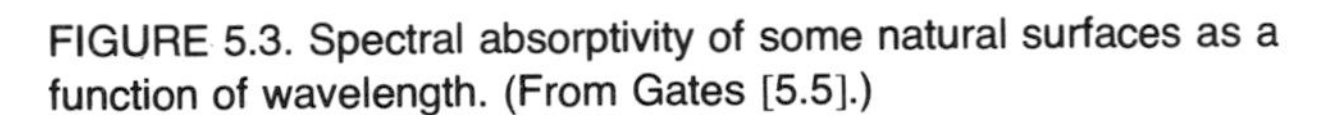

FIGURE 5.3. Spectral absorptivity of some natural surfaces as a function of wavelength. (From Gates [5.5].)

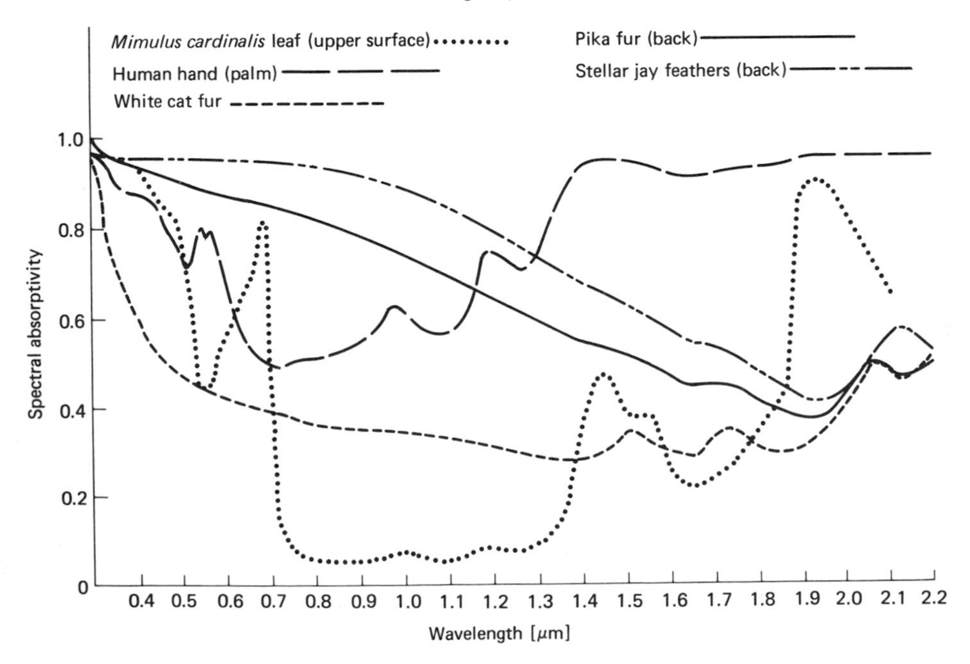

surement at the time the energy budget is being determined. The second type of study is an effort to model the behavior of parts of an ecosystem. The detail is not as important as the ability to obtain reasonable estimates of average radiant flux densities. For the latter studies ecologists often rely on empirical relationships to obtain the estimates they need. Some of these empirical relationships will be presented in the material to follow. Though they can be counted on to give reasonable estimates (± 10 percent) of average flux densities under most conditions, it should be recognized that they are usually not adequate as substitutes for careful field measurements of radiant fluxes for detailed energy budget studies.

Streams of short-wave radiation received by an organism come (1) directly from the sun, (2) from solar radiation scattered by sky and clouds, and (3) from solar radiation reflected from or transmitted by terrestrial objects. It is necessary to separate short-wave radiation into at least this many components because the amount and directional characteristics are different for each. Direct solar radiation is highly directional and irradiance at a surface is determined using Lambert's cosine law (Equation 5.1). Scattered and reflected radiation comes from all directions, and is called diffuse radiation. The diffuse irradiance at a surface is computed by integrating the flux from the surroundings throughout the entire hemisphere around the surface, with each contribution multiplied by the cosine of the angle between the direction of the flux and a normal to the surface. This process can be simplified considerably by making several assumptions. We will assume that diffuse radiation is isotropic (not dependent on direction) for an upper (sky) hemisphere and for a lower (ground) hemisphere. Actually, sky radiance is higher near the horizon than directly overhead on clear days, and higher overhead than near the horizon on overcast days. Also, the sky radiance is considerably higher near the sun than in other parts of the sky due to forward scattering of radiation from the solar beam. These considerations, though sometimes important, can be ignored for most ecological studies. Likewise, the spatial distribution of diffuse flux from the ground can be highly variable, depending on the distribution of shadows and the reflectivity of the substrate, but this too can be averaged to give a reasonable estimate of incoming diffuse radiation for a plant or animal. For surfaces facing up or down, the incoming diffuse irradiance is either sky diffuse irradiance or reflected ground diffuse irradiance. For surfaces that view some sky and some ground, the irradiance can be estimated by summing appropriately weighted contributions from sky and ground. The weighting can be estimated accurately enough for most purposes by determining the approximate fraction of sky and ground viewed by the surface. A quantitative approach, which can be used for more precise estimates, is given by Monteith [5.9]. For detailed field observations, a fisheye camera can be used to record the distribution of diffuse radiation sources for later laboratory analysis.

Estimating Direct and Diffuse Short-wave Irradiance

Computation of the short-wave radiant energy budget of an organism requires estimates of flux densities for at least three radiation streams: direct solar irradiance on a surface perpendicular to the beam (S_p), diffuse sky irradiance (S_d), and reflected radiation from the ground. Reflected radiation is estimated from the product of the average surface reflectivity and the total short-wave irradiance of the surface (S_t). The average surface reflectivity is called albedo. Typical short-wave reflectivities for several surfaces are given in Table 5.1. The values in Table 5.1 should be used with some caution, since all are influenced by amount of cover, color of soil or vegetation, and sun elevation angle. Tall canopies and water surfaces have reflectivities that are particularly elevation angle dependent. The values in Table 5.1 are for high elevation angles, and should be used with caution when the sun angle is below 30 degrees or so.

The total irradiance on a horizontal surface, S_t (which is used to compute the reflected short-wave flux), is the sum of the direct irradiance on a horizontal surface (S_b) and the diffuse sky irradiance:

$$S_t = S_b + S_d. \tag{5.6}$$

The direct irradiance for a horizontal surface is

$$S_b = S_p \sin \phi \tag{5.7}$$

where ϕ is the sun elevation angle from the horizon. Elevation angle is more convenient to measure than the zenith angle

Table 5.1 Short-wave reflectivity (albedo) of soils and vegetation

Surface	Reflectivity	Surface	Reflectivity
Grass	0.24	Deciduous woodland	0.18
Wheat	0.26	Coniferous woodland	0.16
Maize	0.22	Swamp forest	0.12
Pineapple	0.15	Open water	0.05
Sugar cane	0.15	Dry soil (light color)	0.32

required for Equation 5.1, and since they are complimentary angles, $\cos\theta = \sin\phi$. The angle ϕ is easily found in the field by measuring the height of some object and the length of shadow it casts on a horizontal surface, and determining the arctangent of the quotient. For modeling purposes, ϕ can be determined from

$$\sin\phi = \sin\lambda \sin\delta + \cos\lambda \cos\delta \cos 15(t - t_o) \quad (5.8)$$

where λ is the latitude, δ is the solar declination corresponding to the time of observation (Table 5.2), t is time of day in hours, and t_o is the time of solar noon. All angles are in degrees.

Though a number of models are available for estimating clear sky S_p and S_d with considerable accuracy [5.8], they require data that are not generally available to the ecologist without special measurements, and are quite complicated to use. We will use the simpler model given by List [5.7]. We expect S_p to be a function of the distance traveled by the solar beam through the atmosphere, the transmissivity of the atmosphere, and the incident flux density. A simple expression combining these factors is

$$S_p = a^m S_{po} \quad (5.9)$$

where S_{po} is the extraterrestrial flux density normal to the solar beam (1.36 kW/m^2), a is an atmospheric transmission coefficient, and m is the optical airmass number, the ratio of slant-path length through the atmosphere to zenith path length. For elevation angles greater than 10 degrees, refraction effects in the atmosphere are negligible, and m is given by

$$m = (P/P_o)/\sin\phi. \quad (5.10)$$

The ratio P/P_o is atmospheric pressure at the observation site divided by sea level atmospheric pressure, and corrects for altitude effects.

The transmission coefficient varies from around 0.9 for a very clear atmosphere, to around 0.6 for a hazy or smoggy atmosphere. A typical value for clear days would be around 0.84.

As a rough estimate of S_d, List [5.7] suggests using half the

Table 5.2 Solar declination angles (in degrees) on the first day of each month

Jan.	−23.1	Apr.	+4.1	July	+23.2	Oct.	−2.8
Feb.	−17.3	May	+14.8	Aug.	+18.3	Nov.	−14.1
Mar.	−8.0	June	+21.9	Sept	+8.6	Dec.	−21.6

difference between the irradiance on a horizontal surface below and above the atmosphere. Thus,

$$S_d = 0.5\ S_{po}(1 - a^m) \sin \phi. \tag{5.11}$$

By trying various values of ϕ in Equation 5.11, one can quickly verify that S_d is an almost constant fraction of S_{po} over most of the range of elevation angles. Thus, we can estimate the clear-sky diffuse irradiance at any time of the day by evaluating Equation 5.11 at $\phi = 90$ degrees. Thus at sea level we expect S_d to range from 100 to 200 W/m² depending on atmospheric turbidity, and to be less than this value at high altitudes.

When clouds obscure the sun, $S_t = S_d$, since there is no direct radiation component. Empirical transmission coefficients have been worked out for various cloud types and used to determine the short-wave irradiance under clouds. Total short-wave irradiance is shown in Figure 5.4 as a function of solar elevation angle for various cloud types.

Atmospheric Long-wave Radiation

Long-wave radiation is emitted and absorbed in a clear atmosphere mainly by water vapor and CO_2, with a narrow O_3 absorption band. Infrared radiation is absorbed or emitted as a result of changes in the vibrational and rotational energy levels in the molecules. Water vapor, CO_2, and O_3 are the only

FIGURE 5.4. Short wave irradiance under cloud cover for various cloud types and solar elevation angles. (After List [5.7].)

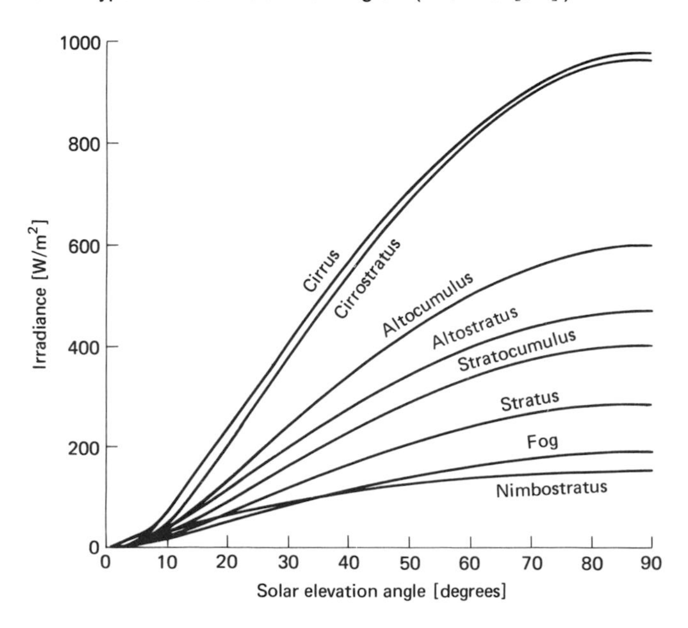

common atmospheric constituents with energy levels that are excited by long-wave radiation. An atmospheric emittance spectrum is shown in Figure 5.5 along with a 288 K blackbody spectrum. Contributions by various atmospheric constituents are indicated. An important feature of this spectrum is the "window" between 8 and 13 μm. This coincides with the blackbody emission peak for the Earth at 288 K, so a rather large fraction of the long-wave radiation emitted by the Earth is lost to space through this "window" when the sky is clear.

The long-wave irradiance from the atmosphere can be determined using Equation 5.4 if one knows the atmospheric emissivity. Several empirical formulae are available for computing estimates of clear sky emissivity. One with reasonable theoretical justification is (Brutsaert [5.1])

$$\epsilon_A = 0.58\ \rho_{va}^{1/7} \tag{5.12}$$

where ρ_{va} is the vapor density measured at 1–2 m height. The reasoning behind this formula is that atmospheric long-wave radiation is primarily a function of the water vapor concentration in the first few kilometers of the atmosphere, and is most strongly dependent on the vapor concentration in the first few hundred meters. Thus a measurement of vapor concentration at 1–2 m height, combined with estimates of vapor and temperature profiles to 5 km, can be used to estimate emissivity.

Since vapor concentration in the atmosphere is correlated

FIGURE 5.5. Emittance spectrum for a 288 K blackbody (upper line) and for the Earth's atmosphere. Atmospheric gases principally responsible for each of the emission bands are indicated. (After Gates [5.4].)

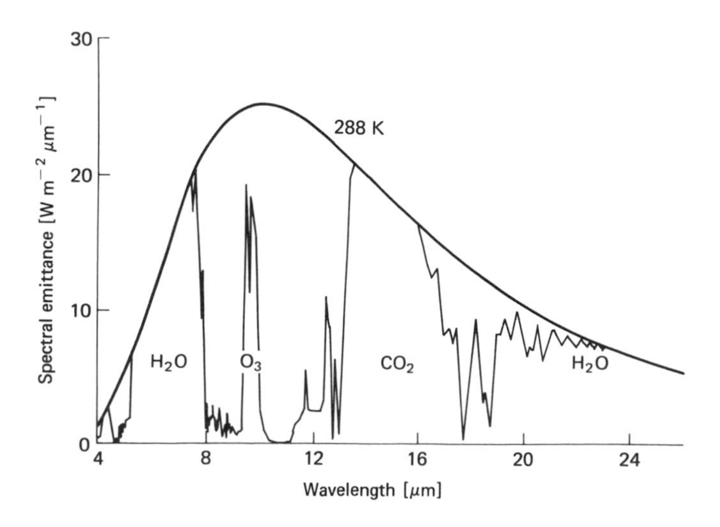

with temperature, correlations have also been made between temperature at 1–2 m height and clear sky emissivity. For temperatures above freezing, the relationship

$$\epsilon_A = 0.72 + 0.005\ T_a \tag{5.13}$$

gives a reasonable fit to available data (T_a is Celsius temperature). Table A.3 gives values for ϵ_A as a function of temperature using this equation. The Idso–Jackson equation [5.6], on which Equation 5.13 is based, gives emissivities at temperatures below freezing, but these estimates may be in error. The low-temperature portion of the equation was fit to a single set of low-temperature data, and more recent data [5.3] shows poor agreement with the Idso–Jackson equation at low temperatures.

Clouds have an emissivity of one, so when clouds are present, atmospheric emissivity is higher than for a clear sky. The atmospheric emittance on cloudy days can be estimated by adding the energy emitted by the clear portions of the sky to the energy emitted by the clouds. The atmospheric emissivity for cloudy days is therefore

$$\epsilon_{Ac} = \epsilon_A + C\ \left(1 - \epsilon_A - \frac{4\Delta T}{T_a}\right) \tag{5.14}$$

with ϵ_A given by Equation 5.13, C the fraction of the sky covered by clouds, and ΔT the difference between T_a and cloud base temperature. Typically ΔT is around 2 K [5.9]. We see that Equation 5.14 predicts $\epsilon_{Ac} = \epsilon_A$ when $C = 0$, and $\epsilon_{Ac} = 1 - 4\Delta T/T_a$ when $C = 1$, as we would expect. Estimates of ϵ_{Ac} are relatively insensitive to errors in estimating C and ΔT since the range of e_{Ac} is relatively small. Estimates accurate to at least ± 10 percent should be possible with very crude estimates of C and ΔT.

Radiant Energy Budgets

Once each of the streams of radiant energy to or from an organism can be estimated or measured, the next step is to determine the radiant energy budget for the organism. The net radiation for a surface is the algebraic sum of all incoming and outgoing streams of radiation. Thus net radiant flux density (per unit projected area) for a small, flat object suspended horizontally above the ground is

$$R_n = 2\ L_{oe} - a_s S_t(1 + \alpha) - a_L(L_{iu} + L_{id}) \tag{5.15}$$

where L_{oe} is the long-wave emittance for each side of the object, L_{iu} and L_{id} are the long-wave irradiances for the up- and down-facing surfaces, α is the albedo of the surface below the object, and a_s and a_L are short- and long-wave absorptivities. Fluxes are taken as positive in the direction away from the heat exchange surface to be consistent with definitions in Chapters 2, 3, and 4.

To illustrate the use of the foregoing equations, let us determine the radiant energy budget of a single leaf suspended horizontally above a grass surface at noon on a clear day, June 1, at 40 degrees north latitude (Figure 5.6). Short-wave absorptivity for the leaf is 0.4 and long-wave absorptivity and emissivity are 0.95. Albedo for the grass is 0.24 (Table 5.2) and emissivity is 0.97. Leaf temperature is 30°C and grass temperature is 40°C. Air temperature is 25°C. From Equation 5.8, the elevation angle of the sun is 71.9 degrees. This gives a sea level air mass of 1.05 (Equation 5.10), and, assuming an atmospheric transmission coefficient of 0.84, a value of S_p of 1132 W/m² (Equation 5.9). The direct irradiance on a horizontal surface is S_b = 1076 W/m² (Equation 5.7). For an atmospheric transmission coefficient of 0.84, the diffuse flux is 8 percent of S_{po} (Equation 5.11), so S_d = 109 W/m². The total short-wave irradiance on a horizontal surface is therefore S_t = 1076 + 109 = 1185 W/m² (Equation 5.6). The long-wave irradiances and emittances are easily found from Table A.3 or Equations 5.4 and 5.13 They are L_{oe} = 0.95 × 479 = 455 W/m², L_{id} = 0.97 × 545 = 529 W/m², and L_{iu} = 0.85 × 448 = 379 W/m². The net radiation for the leaf is R_n = 2 × 455 − 0.4 × 1185 × 1.24 − 0.95 × (379 − 529) = −540 W/m² (Equation 5.15). Again the negative sign means that the net flux is directed toward the leaf.

For surfaces that are not horizontal, the radiation budget equation is more complicated because the short-wave irradiance must be separated into direct and diffuse components, and the Lambert cosine law applied to determine the direct

FIGURE 5.6. Streams of radiant energy toward and away from a leaf.

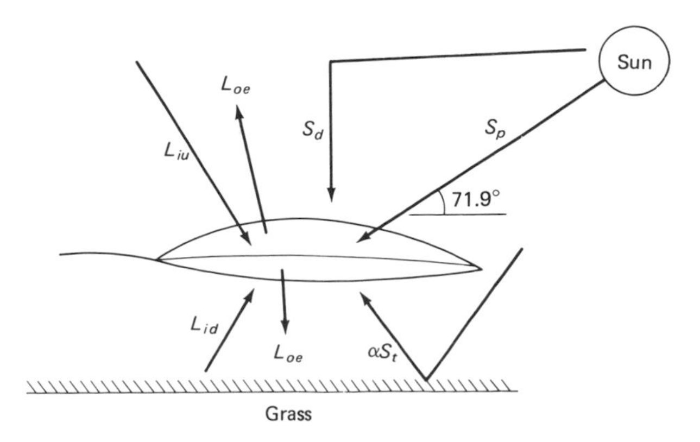

short-wave component using the angle between a normal to the surface and the solar beam. We will leave this exercise to later chapters where we apply radiation budgets to animals and leaves.

References

5.1 Brutsaert, W. (1975) On a derivable formula for long-wave radiation from clear skies. *Water Resour. Res. 11:* 742–744.

5.2 Coulson, K. L (1975) *Solar and Terrestrial Radiation*. New York: Academic Press.

5.3 Enz, J. W., J. C. Klink, and D. G Baker (1975) Solar radiation effects on pyrgeometer performance. *J. Appl. Meteor. 14*:1297–1302.

5.4 Gates, D. M. (1962) *Energy Exchange in the Biosphere*. New York: Harper and Row.

5.5 Gates, D. M. (1975) Introduction: Biophysical ecology. *Perspectives of Biophysical Ecology* (D. M. Gates and R. B Schmerl, eds.). New York: Springer Verlag.

5.6 Idso, S. B. and R. D Jackson (1969) Thermal radiation from the atmosphere. *J. Geophys. Res. 74*:5397–5403.

5.7 List, R. J. (1971) *Smithsonian Meterological Tables,* 6th Ed. Washington, D.C.: Smithsonian Institution Press.

5.8 McCullough, E. C. and W. P. Porter (1971) Computing clear day solar radiation spectra for the terrestrial ecological environment. *Ecology 52*:1008–1015.

5.9 Monteith, J. L. (1973) *Principles of Environmental Physics*. New York: American Elsevier.

5.10 Munn, R. E. (1966) *Descriptive Micrometerology*. New York: Academic Press.

GENERAL REFERENCE

Gates, D. M. (1965) Radiant energy, its receipt and disposal. *Meteor. Monogr. 6*:1–26.

Problems

5.1 The median wavelength for solar radiation in the 0.3–3 μm waveband is approximately 0.7 μm. If the irradiance is 1 kW/m^2, what is the photon flux?

5.2 If your average surface temperature is 30°C, what is your radiant emittance? If the average wall temperature of the room is 20°C, what is the average long-wave irradiance at your body surface? What is your net long-wave radiation (outgoing − incoming)? Assume $\epsilon_{you} = 0.97$, $\epsilon_{room} = 1$.

5.3 Compare clear night sky, completely overcast cloudy night sky, and ground emittance for $T_a = T_{soil} = 20°C$. What is the incoming long-wave radiant flux density for an organism (average of sky and ground) under a clear sky and under a cloudy sky?

5.4 Show a complete radiant energy budget for your hand suspended horizontally over a dry soil surface on a clear day. Assume $T_a = 20°C$, $T_s = 35°C$, $T_{soil} = 45°C$, $\phi = 40$ degrees, $a_s = 0.65$. Estimate values not given and state assumptions involved in making estimates.

6 Heat, Mass, and Momentum Transfer

Life depends on heat and mass transfer between organisms and their surroundings. Such processes as carbon dioxide exchange between leaves and the atmosphere, oxygen uptake by microorganisms, oxygen and carbon dioxide exchange in the lungs of animals, or convective heat loss from the surfaces of animal coats are fundamental to the existence of living organisms. A thorough understanding of these exchange processes is therefore a necessary part of the study of physical ecology. In this chapter we will first discuss molecular diffusion. It is by this process that heat and mass are transported in still air or water, as they are in parts of the lungs of animals, in soils, and in the substomatal cavities of leaves. Molecular diffusion is also important in convective heat and mass transfer between surfaces and fluids flowing over them since a thin boundary layer is always formed near the surface through which transport is by diffusion. After diffusion processes are discussed, we will then present convective heat and mass transfer theory as it applies to fluids moving over plates, cylinders, and spheres. Finally, we will discuss momentum exchange and the force of moving fluids on objects in them.

The description we will use for heat and mass transport was outlined in Equations 1.1, 2.1, and 3.1. Heat flux density at the surface of an organism is given by

$$H = \rho c_p \frac{T_s - T_a}{r_H} \tag{6.1}$$

flux density of water vapor by

$$E = \frac{\rho_{vs} - \rho_{va}}{r_v} \tag{6.2}$$

and flux density of other materials such as oxygen, CO_2, and pollutants by

$$F_j = \frac{\rho_{js} - \rho_{ja}}{r_j} \tag{6.3}$$

where the subscripts s and a indicate the value of the variable at the surface and in the surroundings. The density of air is ρ (g/m^3), of vapor, ρ_v, and of other material, ρ_j. The specific heat of air is c_p, and T is temperature. The resistance to transfer of heat or mass from the surface to its surroundings is r (s/m). Until we can predict r from known properties of the surface, the diffusing substance, and the surroundings, these equations are of little value to us except to indicate the nature of the driving forces.

Molecular Diffusion

Analytical descriptions of r can be most easily obtained for conditions where heat or mass are transported by random molecular motion. Fick's law for one-dimensional diffusion of some component, j, in a system is

$$F_j(x) = -D_j \frac{d\rho_j}{dx} \tag{6.4}$$

where D_j is the diffusivity of the component j, and $d\rho_j/dx$ is the concentration gradient. If we assume that all of the material that diffuses across an imaginary boundary at x originated at a surface with area $A(s)$ where the flux density is $F_j(s)$, a constant, then

$$F_j(s) = \frac{A(x)}{A(s)} F_j(x) = -\frac{A(x)}{A(s)} D_j \frac{d\rho_j}{dx} \tag{6.5}$$

where $A(x)$ is the area of the surface at x. Integration of Equation 6.5 and substitution from Equation 6.3 gives

$$r_j = \frac{A(s)}{D_j} \int_s^l \frac{dx}{A(x)}. \tag{6.6}$$

Equation 6.6 can be easily worked out for several simple shapes. For planar diffusion, such as diffusion through long, narrow tubes or from extended surfaces $A(x) = A(s)$ and

$$r_j = \frac{l}{D_j} \tag{6.7}$$

with l being the distance from the source ($s = 0$) to the point at which ρ_{ja} is measured. The decrease in concentration is linear with distance from the surface.

For diffusion from a spherical surface, $A(x) = 4\pi x^2$ (x is the radial distance from the center of the sphere). The resistance at a distance l from the center of the sphere is

$$r_j = \frac{s}{D_j}\left(1 - \frac{s}{l}\right) \tag{6.8}$$

where the radius of the spherical surface is s. In the limit, as l becomes very large $s/l \rightarrow 0$, and $r_j = s/D_j$.

For a cylindrical surface with unit length $A(x) = 2\pi x$ (x is the distance from the axis of the cylinder), and integration of Equation 6.6 gives

$$r_j = \frac{s}{D_j} \ln \frac{l}{s} \tag{6.9}$$

where s is the radius of the cylindrical exchange surface and l is the distance from the cylinder axis to the point of measurement. Equation 6.6 could be integrated for other shapes, if desired.

Before we can use Equations 6.7 to 6.9, we need to have values for D_j. These depend on the properties of the diffusing substance and the medium in which diffusion occurs. Molecular diffusion coefficients for heat, water vapor, oxygen, and CO_2 in air are given in Table A.1. Available data for diffusion in water are given in Table A.2. Diffusivities in air can be corrected for temperature and pressure effects using [6.9]

$$D_j(T,P) = D_j^0 \left(\frac{T}{T_o}\right)^n \left(\frac{100}{P}\right) \tag{6.10}$$

With D_j^0 the diffusivity at sea level (100 kPa) and temperature, T_o, and $n = 2$ for CO_2 and 1.75 for other gases. We will use 20°C (293 K) as our reference for which $D_H^0 = 21.5$ mm²/s and $D_v^0 = 24.2$ mm²/s. Other values are given in the appendix.

Equations 6.7 to 6.9 have been used in numerous biological environment studies. For example, oxygen movement in soil is a diffusion process. Variants of Equation 6.9 have been used to describe oxygen flux to roots and oxygen concentrations at the root surface [6.8]. Equation 6.8 could be used to describe oxygen concentrations within water-saturated soil aggregates which form spherical "anaerobic microsites" in the soil [6.4]. These microsites are said to be sites of ethylene production, and it is apparently critical that oxygen concentration be low there for ethylene to be produced [6.13]. Ethylene is a growth regulator, and when present in the soil in the proper concentrations can control some soil-borne pathogens [6.1]. The movement of ethylene out of the soil is also a diffusion process. The ethylene concentration at any particular time would depend on the rate of production and the resistance to ethylene diffusion. The rate of production apparently depends on the rate of oxygen diffusion to the microsites. Manipulation of ethylene levels in soil to effect control of pathogens would therefore require a good understanding of the physics of the system as well as the biology.

Another problem to which diffusion theory has been applied is the transport of gases through stomates in leaves. The resistance to diffusion within a single stomatal pore is l/D_j (Equation 6.7) where l is the pore depth. To account for nonplanar diffusion just outside the stomatal pore an "end correction" is applied. The total surface resistance is given by

$$r_{js} = \frac{4\left(l + \pi \dfrac{d}{8}\right)}{\pi n d^2 D_j} \tag{6.11}$$

where d is the pore diameter (m) and n is the number of pores per square meter. $4l/\pi n d^2 D$ is just the fraction of a square meter covered by pores multiplied by l/D, and $4\pi d/8\pi n d^2 D$ is the end correction. Equation 6.11 is valid for water vapor, CO_2, or oxygen when the appropriate diffusion coefficient is used. Interactions between diffusing species and convection corrections are important in some studies [6.5], but will not be discussed here.

Convection

In the previous section we discussed transfer of heat and mass by molecular diffusion. The fluid was assumed still or moving with laminar flow over an infinitely long surface so that there were no concentration gradients in the direction of flow. We now want to consider convective transport to or from a small object such as an animal's body or a leaf that is immersed in a fluid such as air. The assumptions made for molecular diffusion do not apply, but Equations 6.1 to 6.3 still apply if the r terms are properly defined. Our task here is to relate the r terms to properties of the fluid and the surface.

A fundamental analysis of convective transport can be extremely complicated and has not been possible for many surface shapes. For this reason an empirical approach has been used. To make the results of empirical studies apply to as many different situations as possible, dimensionless groups of the variables have been formed and correlated empirically. By using these dimensionless groups, the results of a study on, say, heat transfer to water by a l-cm-diameter rod, can be used to calculate heat transfer from a man's arm to air. We would therefore like to relate our r terms to the appropriate dimensionless groups. Once this is done, the appropriate relationships between the dimensionless groups for describing the transport processes can be obtained directly from the engineering literature. The dimensionless groups that we will use are given in Table 6.1. The Reynolds number, Re (Table 6.1) besides being useful for correlation of data on heat and mass transport, also gives an indication of whether the flow is laminar or turbulent. At low Re, viscous forces predominate, and

Table 6.1 Dimensionless groups for heat and mass transfer

Name	Equation	Symbols	Explanation
Reynolds number	$\mathrm{Re} = \frac{ud}{v}$	u = Fluid velocity v = Kinematic viscosity d = Characteristic dimension (diameter of cylinders and spheres, length of plates)	Ratio of inertial forces to viscous forces
Grashof number	$\mathrm{Gr} = \frac{agd^3(T_s - T_a)}{v^2}$	g = Gravitational acceleration (9.8 m/s^2) a = Coefficient of thermal expansion for fluid (a = 1/273 for air) T_s = Surface temperature T_a = Fluid temperature	Ratio of a buoyant force times an inertial force to the square of a viscous force
Nusselt number	$\mathrm{Nu} = \frac{H\,d}{\rho c_p D_H (T_s - T_a)}$	H = Heat flux density ρ = Fluid density c_p = Fluid specific heat D_H = Thermal diffusivity	Ratio of the actual heat flux to that which would result from the same temperature difference applied to a still fluid layer of depth d
Sherwood number	$\mathrm{Sh} = \frac{F_j d}{D_j(\rho_{js} - \rho_{ja})}$	F = Mass flux density D = Molecular diffusion coefficient ρ_{js} = Concentration (g/m^3) at the surface ρ_{ja} = Concentration in bulk fluid	Ratio of actual mass flux density to that which would result from the same concentration applied to a still fluid layer of depth d
Prandtl number	$\mathrm{Pr} = \frac{v}{D_H}$		Ratio of the kinematic viscosity to the thermal diffusivity
Schmidt number	$\mathrm{Sc} = \frac{v}{D_j}$		Ratio of kinematic viscosity to molecular diffusivity

the flow is laminar. At high Re, inertial forces predominate and the flow becomes turbulent. The critical Re at which turbulence starts is around 5×10^5 for a smooth, flat plate under "average" conditions. In the atmosphere, which itself is turbulent, the critical Re may be a factor of 10 lower.

Some idea of usual Reynolds numbers for biological systems is given in the following example:

Example. Determine Re for a man and a bee at windspeeds of 1 m/s (~2 mph) and 10 m/s (~20 mph). Assume the man has a diameter of $d = 0.3$ m and the bee has $d = 0.003$ m.

man, 1 m/s: $$\mathrm{Re} = \frac{0.3\ \mathrm{m} \times 1\ \mathrm{m/s}}{151 \times 10^{-7}\mathrm{m^2/s}} = 2 \times 10^4$$

man, 10 m/s: $$\mathrm{Re} = \frac{0.3\ \mathrm{m} \times 10\ \mathrm{m/s}}{151 \times 10^{-7}\mathrm{m^2/s}} = 2 \times 10^5$$

bee, 1 m/s: $$\mathrm{Re} = \frac{0.003\ \mathrm{m} \times 1\ \mathrm{m/s}}{151 \times 10^{-7}\mathrm{m^2/s}} = 2 \times 10^2$$

bee, 10 m/s: $$\mathrm{Re} = \frac{0.003\ \mathrm{m} \times 10\ \mathrm{m/s}}{151 \times 10^{-7}\mathrm{m^2/s}} = 2 \times 10^3.$$

Resistances to Heat and Mass Transfer in Laminar Forced Convection

Forced convection refers to the condition in which a fluid is moved past a surface by some external force (the analysis would be the same for a surface moving through a stationary fluid). Free convection, on the other hand, refers to fluid motion brought about by density gradients in the fluid as it is heated or cooled by the exchange surface. The resistance to heat transfer for a surface with forced convection is obtained by combining Equation 6.1 with the definition of the Nusselt number from Table 6.1 to give

$$r_H = \frac{d}{D_H \mathrm{Nu}}. \tag{6.12}$$

We require an estimate of Nu to obtain r_H. For a flat plate in laminar flow, heat transfer theory provides the following relationship [6.7]:

$$\mathrm{Nu} = 0.66\ \mathrm{Re}^{1/2}\mathrm{Pr}^{1/3}. \tag{6.13}$$

Equation 6.12 therefore becomes

$$r_H = \frac{1.5\ d}{D_H \mathrm{Re}^{1/2}\mathrm{Pr}^{1/3}}. \tag{6.14}$$

For air at 20°C and 100 kPa pressure, substitution from Table 6.1 for Re gives

$$r_{Ha} = 307 \left(\frac{d}{u}\right)^{1/2} \tag{6.15}$$

when d is in m and u is in m/s, r_H is in s/m.

Equation 6.15 was derived only for laminar flow over flat plates. It gives the resistance to heat transfer from *one* side of the plate. Correlations similar to Equation 6.13 for cylinders in cross-flow and spheres have been determined experimentally [6.3, 6.7, 6.9], but the range of variation between correlations from different studies is great. Equation 6.13 appears to be a good average of all sphere and cylinder correlations for usual Reynolds numbers found in nature. We will therefore use Equation 6.15 for plates, cylinders, and spheres, with d being the diameter of a cylinder (in cross-flow) or sphere or the length of a plate (or cylinder in longitudinal flow) in the direction of flow.

By combining Equation 6.2 or 6.3 with the definition of the Sherwood number (Table 6.1) we obtain

$$r_j = \frac{d}{D_j \mathrm{Sh}} . \tag{6.16}$$

For a flat plate in laminar flow, mass transfer theory provides the following equation [6.7]

$$\mathrm{Sh} = 0.66\ \mathrm{Re}^{1/2}\mathrm{Sc}^{1/3} \tag{6.17}$$

From this, the resistance equation is

$$r_j = \frac{1.5\ d}{D_j\, \mathrm{Re}^{1/2}\mathrm{Sc}^{1/3}} \tag{6.18}$$

Substitution of the appropriate variables for Sc and Re from Table 6.1 gives the following relationships for air at 100 kPa and 20°C:

$$r_{va} = 283 \left(\frac{d}{u}\right)^{1/2} \tag{6.19}$$

$$r_{ca} = 395 \left(\frac{d}{u}\right)^{1/2} \tag{6.20}$$

$$r_{oa} = 319 \left(\frac{d}{u}\right)^{1/2} \tag{6.21}$$

where r_c is for CO_2 and r_o is for oxygen. These equations are also approximately correct for cylinders and spheres. Similar equations for transport of other materials or transport in fluids other than air can be derived from Equations 6.14 and 6.18 if the laminar flow and steady-state conditions are met.

Example. Find the flux densities of heat, CO_2, water vapor, and oxygen for a leaf under the following conditions:

	Air	Leaf
T	20°C	30°C
u	3 m/s	0
ρ_v	10 g/m³	30 g/m³
ρ_c	0.55 g/m³	0.12 g/m³
ρ_o	270.00 g/m³	270.25 g/m³

Assume the leaf has a characteristic dimension in the direction of wind flow of $d = 0.042$ m and that surface resistances (stomatal) to water vapor, CO_2, and O_2 transfer are $r_{vs} = 200$ s/m, $r_{cs} = 328$ s/m, and $r_{os} = 240$ s/m.

Solution. The Reynolds number is Re $= ud/v = 3 \text{ m s}^{-1} \times 0.042 \text{ m}/151 \times 10^{-7} \text{ m}^2\text{s}^{-1} = 8.4 \times 10^3$. The flow apparently is laminar. The resistances all involve

$$\left(\frac{d}{u}\right)^{1/2} = \left(\frac{0.042}{3}\right)^{1/2} = 0.118.$$

The resistances are

$$r_{va} = 33 \text{ s/m} \qquad r_{Ha} = 36 \text{ s/m},$$
$$r_{ca} = 47 \text{ s/m} \qquad r_{oa} = 38 \text{ s/m}.$$

The fluxes are

$$H = \frac{\rho c_p(T_s - T_a)}{r_H} = \frac{1.2 \times 10^3 \text{ J m}^{-3} \text{ °C}^{-1} (30\text{°C} - 20\text{°C})}{36 \text{ s/m}}$$

$$= 333 \text{ W/m}^2$$

$$E = \frac{\rho_{vs} - \rho_{va}}{r_v} = \frac{(30 - 10) \text{ g/m}^3}{(200 + 33) \text{ s/m}} = 0.086 \text{ g m}^{-2}\text{s}^{-1}$$

$$F_c = \frac{\rho_{cs} - \rho_{ca}}{r_c}$$

$$= \frac{(0.12 - 0.55) \text{ g/m}^3}{(328 + 47) \text{ s/m}} = -1.15 \times 10^{-3} \text{ g m}^{-2}\text{s}^{-1}$$

$$F_o = \frac{\rho_{os} - \rho_{oa}}{r_o}$$

$$= \frac{(270.25 - 270.00)\ \text{g/m}^3}{(240 + 38)\ \text{s/m}} = 9 \times 10^{-4}\ \text{g m}^{-2}\text{s}^{-1}.$$

Note that the resistance for all but heat transfer are the sums of surface and boundary layer resistances.

Free Convection

Transport by free convection occurs whenever a body at one temperature is placed in a fluid at a higher or lower temperature. The heat transfer between the body and the fluid causes density gradients in the fluid, and these density gradients cause the fluid to mix. The transfer processes are similar to those in forced convection except that, where fluid velocity increases with distance from the surface in forced convection, it increases, then decreases again in free convection. This is shown in Figure 6.1. For heat transfer, the observations are correlated using the dimensionless groups Nu, Pr, and Gr (Table 6.1). For laminar free convection, an expression that is adequate for cylinders (horizontal or vertical), spheres, vertical flat plates, and heated flat surfaces facing up or cooled surfaces facing down is [6.7]

$$\text{Nu} = 0.54(\text{GrPr})^{1/4}. \tag{6.22}$$

For cooled flat surfaces facing up or heated surfaces facing down the heat transfer is only about half as efficient [6.7], so

FIGURE 6.1. Comparison of fluid velocity distributions in free and forced convection.

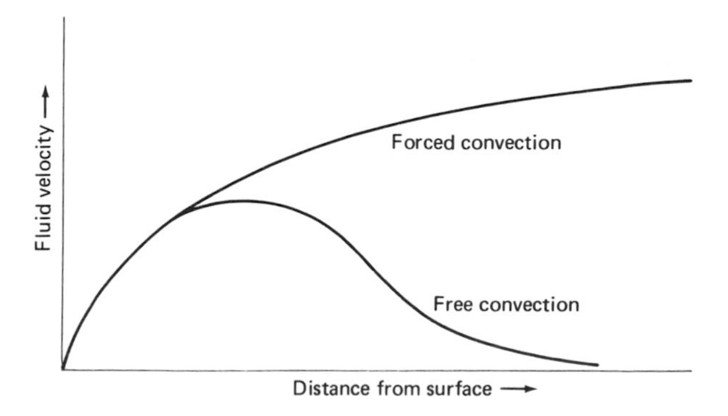

$$\mathrm{Nu} = 0.26(\mathrm{GrPr})^{1/4}. \tag{6.23}$$

Combining Equation 6.12 with Equation 6.22, we can express r_H as

$$r_H = \frac{d}{D_H \mathrm{Nu}} = \frac{d}{0.54 D_H (\mathrm{GrPr})^{1/4}}. \tag{6.24}$$

For air at 20°C and 100 kPa, and using the definition of Gr from Table 6.1, we can write

$$r_H = 840 \left(\frac{d}{T_s - T_a}\right)^{1/4}. \tag{6.25}$$

when r_H is in s/m, d is in m, and T is in °C.

For mass transfer by free convection,

$$r_j = \frac{d}{D_j \mathrm{Sh}} = \frac{d}{0.54 D_j (\mathrm{GrSc})^{1/4}}. \tag{6.26}$$

The free convection resistances are 0.92, 1.34, and 1.05 times the heat transfer resistance (Equation 6.25) for water vapor, CO_2 and O_2, respectively.

When water vapor is transported density gradients develop just as with heated air, since moist air is less dense than dry air. This is conveniently accounted for by using the virtual temperature in Equation 6.25 rather than the actual temperature. The virtual temperature is the temperature at which dry air would need to be to make its density the same as moist air, with a given vapor density.

It is approximated by

$$T_v = T\left(1 + \frac{0.61\,\rho_v}{\rho}\right).$$

The temperature difference for Equation 6.25 would therefore be

$$T_{vs} - T_{va} = T_s - T_a + \frac{0.61(\rho_{vs} T_s - \rho_{va} T_a)}{\rho}. \tag{6.27}$$

In wet systems with small $T_s - T_a$, the correction term in Equation 6.27 may be several times larger than $T_s - T_a$, so the contribution can be important.

Combined Forced and Free Convection

Almost all convective heat transfer processes in nature involve both forced and free convection. Usually one or the other process dominates, and the resistance used is that calculated for the dominant process. The criterion normally used to determine which process is dominant is to determine the ratio

Gr/Re². If this ratio is small, forced convection dominates. When the ratio is large, the opposite is true. When the ratio is near one, both forced and free convection must be considered, and the value of r depends on whether the direction of the forced convection flow is such that it enhances or diminishes free convection. Additional detail on the mixed regime can be found in Kreith [6.7].

Example. Find Gr for the previous example, and compute the ratio Gr/Re².

$$\mathrm{Gr} = \frac{a\ g\ d^3(T_s - T_a)}{v^2}$$

$$= \frac{(9.8\ \mathrm{m/s^2}) \times (0.042\ \mathrm{m})^3 \times (30°\mathrm{C} - 20°\mathrm{C})}{273\,\mathrm{K} \times (151 \times 10^{-7}\ \mathrm{m^2/s})^2}$$

$$= 1.2 \times 10^5$$

$$\frac{\mathrm{Gr}}{\mathrm{Re}^2} = \frac{1.2 \times 10^5}{(8.4 \times 10^3)^2} = 1.7 \times 10^{-3}.$$

Forced convection obviously dominates.

Application in Nature

To apply the transport equations to real situations, we need to answer three questions. These are: (a) What correction needs to be applied to the flat plate equations to account for the effect of shapes of natural surfaces on r? (b) When transfer is from closely spaced surface elements, what is the correct r to use? (c) What is the effect of free stream turbulence on r?

Question (a) can be answered by considering the transport for individual narrow strips of a surface and averaging these to get the total transport. A new value for d is derived which can be used to get a correct value for r. Figure 6.2 shows a possible leaf outline and indicates the length and width of strips by $d(x)$ and dx. The total length of the leaf is W. The appropriate value for d in forced convection is given by [6.9]

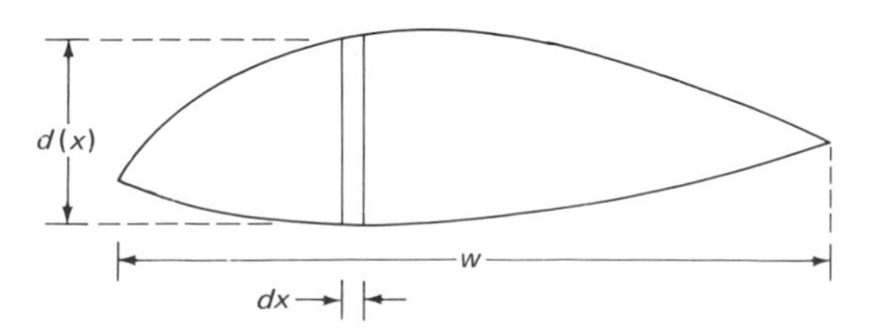

FIGURE 6.2. Leaf outline showing the definitions of variables for Equation 6.28.

$$d = \left\{ \frac{\int_0^w d(x)\,dx}{\int_0^w [d(x)]^{1/2} dx} \right\}^2. \tag{6.28}$$

Equation 6.28 gives the width of a rectangular plate of length, w, which has the same average transport properties as the leaf. In practice, the *average* leaf width is usually a satisfactory estimate of d. For many leaf shapes, d can be taken as 0.7 times the maximum leaf dimension in the direction of flow, since the leaf area is about 0.7 × length × width [6.2].

A quantitative answer for question (b) is harder to provide. The few studies that have been done indicate that r may increase by 50 percent for closely spaced elements. As a rule, if the elements in an assemblage (leaflets of a leaf, etc.) are spaced farther than one or two characteristic dimensions from each other, they probably have negligible mutual interference. Closer spacing will cause r to increase.

Several studies have shown that the boundary layer resistance in turbulent flow is smaller than is measured in wind tunnel tests or predicted from theory. This is because turbulence in the air tends to reduce the depth of the laminar boundary layer around the surface. Resistances with free-stream turbulence are a function of turbulence intensity, characteristic dimension, and eddy size. For typical outdoor turbulence intensities, the resistance is around 70 percent of the values predicted by Equations 6.15 and 6.19 to 6.21 [6.6, 6.10].

Momentum Transport

When a fluid moves over a surface, the fluid exerts a frictional force on the surface in the direction of flow. The surface exerts an equal and opposite retarding force on the fluid. The fluid velocity at the surface is zero, and increases with distance from the surface. If a thin, flat plate is placed in a moving fluid in such a way that the fluid flows smoothly over the plate, the momentum flux density to one side of the plate is given by

$$\tau_s = 0.664\ \rho u^2 \mathrm{Re}^{-1/2}. \tag{6.29}$$

This flux density is called skin friction. Momentum flux could also be expressed as a difference in momentum concentration between the free stream and the surface, divided by a resistance:

$$\tau = \rho u / r_M \tag{6.30}$$

where r_M is defined as:

$$r_M = 1.5\ \mathrm{Re}^{1/2}/u.$$

For air at 20°C and 100 kPa,

$$r_M = 388\left(\frac{d}{u}\right)^{1/2}. \tag{6.31}$$

A fluid moving past a bluff body exerts a force on the body due to changes in direction of fluid flow in addition to the force of skin friction. At Reynolds numbers above about 10, this force, called *form drag,* is much greater than skin friction. At Re > 100, even streamlined forms develop significant form drag. We are familiar with form drag as the force of wind or moving water on us or the bending of wheat or trees by the wind. An understanding of form drag should allow us to calculate the force of the wind on trees or crop plants, and thus allow prediction of windthrow and lodging. We could also calculate the work required for an animal to move through water or air.

The average force on a body due to form drag per unit cross-section area perpendicular to flow can be calculated from

$$\tau_f = \frac{\text{Force}}{\text{Area}} = C_d \rho u^2 \tag{6.32}$$

where C_d is the drag coefficient, ρ is the fluid density, and u is the fluid velocity. Note that Equation 6.32 looks similar to 6.29 (with $C_d = 0.664\ \text{Re}^{-1/2}$), but remember that Equation 6.29 is for the force on an area tangential to flow while Equation 6.32 is for the force on an area perpendicular to flow.

Table 6.2 gives drag coefficients of several common shapes for a range of Re and shows how the area for Equation 6.32 is determined. At Re lower than the values shown, C_d increases and approaches the value predicted by Equation 6.29. At Re higher than the values shown, C_d decreases abruptly because of the onset of turbulence.

Example. Find the total drag force on a fish and the energy required to swim at a rate of 1 m/s. Assume the average diameter is 0.05 m, and that the hydrodynamic properties of the fish are similar to those of an airship hull. The Reynolds number for the fish in water is

$$\text{Re} = \frac{ud}{v} = \frac{1 \times 0.05}{10.1 \times 10^{-7}} = 5 \times 10^4.$$

Table 6.2 indicates that a value of $C_d = 0.035$ can be used for this Reynolds number. The total force on the fish is

$$\tau_f A = C_d A \rho u^2$$

$$= 0.035 \times \pi \times \frac{(0.05 \text{ m})^2}{4} \times 10^3 \text{ kg/m}^3 \times (1 \text{ m/s})^2$$

$$= 0.07 \text{ N}.$$

Table 6.2 Typical drag coefficients for various body forms [6.12]

Form	d	A	$\frac{L}{d}$	Re	C_d
Circular disk	Fluid flow → ; d	$\frac{\pi d^2}{4}$		$> 10^3$	0.56
Rectangular plate, L = length, d = width	d	Ld	1	$> 10^3$	0.58
			5		0.60
			20		0.75
			∞		0.95
Circular cylinder (axis || to flow.)	d; L	$\frac{\pi d^2}{4}$	0	$> 10^3$	0.56
			1		0.46
			2		0.43
			4		0.44
			7		0.50
Circular cylinder (axis ⊥ to flow), L = length	d	Ld	1	10^2–10^5	0.32
			5		0.37
			20		0.45
			∞		0.60
Streamlined foil, L = length	d	Ld	∞	$> 4 \times 10^4$	0.035
Sphere	d	$\frac{\pi d^2}{4}$		10^3–10^5	0.25
Hemisphere: hollow upstream	d	$\frac{\pi d^2}{4}$		$> 10^3$	0.67
Hemisphere: hollow downstream	d	$\frac{\pi d^2}{4}$		$> 10^3$	0.17
Airship hull	d	$\frac{\pi d^2}{4}$		10^4–10^6	0.035

The power required is the force multiplied by the velocity or $u\tau A$ = 1 m/s × 0.07 N = 0.07 W. To get some feeling for what this means, if the fish were powered by a flashlight battery, he could swim for about a day at this rate before the battery ran down. Another way to look at it is to ask how long a kilogram of food would last the fish. The caloric content of glucose is 15.7 MJ/kg. If we assume the fish converts food to work with an efficiency of around 1/3, then swimming will require 3 × 0.07 W = 0.21 J/s. A kilogram of glucose would therefore supply enough energy to swim for 15.7×10^6 J kg^{-1}/0.21 J s^{-1} = 7.5×10^7s, or about 863 days.

References

6.1 Baker, K. F. and R. J. Cook (1974) *Biological Control of Plant Pathogens*. San Francisco: W. H. Freeman.

6.2 Cowan, I. R. (1972) Mass and heat transfer in laminar boundary layers with particular reference to assimilation and transpiration in leaves. *Agric. Meteor. 10*:311–329.

6.3 Eckert, E. R. G. and R. M. Drake (1972) *Analysis of Heat and Mass Transfer* New York: McGraw-Hill.

6.4 Griffin, D. M. (1972) *Ecology of Soil Fungi*. Syracuse, N.Y.: Syracuse University Press.

6.5 Jarman, P. D. (1974) The diffusion of carbon dioxide and water vapor through stomata. *J. Exp. Bot. 25*:927–936.

6.6 Kowalski, G. J. and J. W. Mitchell (1975) Heat transfer from spheres in the naturally turbulent, outdoor environment. *Amer. Soc. Mech. Eng.* Paper No. 75-WA/HT-57.

6.7 Kreith, F. (1965) *Principles of Heat Transfer*. Scranton, Pa.: International Textbook Co.

6.8 Lemon, E. R. and C. L. Wiegand (1962) Soil aeration and plant root relations II. Root respiration. *Agron. J. 54*:171–175.

6.9 Monteith, J. L. (1973) *Principles of Environmental Physics*. New York: American Elsevier.

6.10 Nobel, P. S. (1974) Boundary layers of air adjacent to cylinders. *Plant Physiol. 54*:177–181.

6.11 Parkhurst, D. F., P. R. Duncan, D. M. Gates, and F. Kreith (1968) Wind tunnel modelling of convection of heat between air and broad leaves of plants. *Agric. Meteor. 5*:33.

6.12 Rouse, H. (1946) *Elementary Mechanics of Fluids*. New York: John Wiley.

6.13 Smith, A. M. and R. J. Cook (1974) Implication of ethylene production by bacteria for biological balance of soil. *Nature 252*:703–705.

Problems

6.1 If the diffusivity for oxygen in soil is given by $D_{soil} = 0.7\, x_a\, D_{air}$, where x_a is the volume fraction of soil filled with air (air-filled porosity), at what porosity will the O_2 concentration at the root surface be 50 g/m³ if $F_o = -1.4 \times 10^{-3}$ g m^{-2}s^{-1}, $\rho_{oa} =$ 150 g/m³ at a distance of 10^{-2} m from the surface, and the root radius is 10^{-3} m. Assume $T = 20$°C

6.2 What is the Reynolds number for an elephant in a 10-m/s wind? (Make your own estimate for d.)

6.3 The ratio r_c/r_v is different for transport through stomates (Equation 6.11), which is pure diffusion, and transport by convection in the boundary layer of the leaf (Equation 6.18). Find r_{cs}/r_{vs} and r_{ca}/r_{va}, and suggest a reason for the difference.

6.4 A leaf with maximum width in the direction of wind flow of 5 cm is at 20°C in a 1-m/s wind ($T_a = 15$°C). Find d, Re, Gr, r_{Ha}, and H. Is heat transfer mainly by forced or free convection?

6.5 What is the total force on you when you face into a 20-m/s wind? Use C_d for a cylinder.

7 Animals and Their Environment

The principles discussed thus far become more meaningful as they are applied to problems in nature. The first of these problems we will consider is that of describing the fitness of the physical environment for survival of some animal whose requirements we will specify. Survival of the animal can depend on many factors; we will consider only those related to maintaining body temperature within acceptable limits and those related to maintaining proper body water status. Even these aspects will only be discussed to a limited extent. For example, maintenance of body temperature in homeotherms involves production of metabolic heat. Stored chemical energy from the animal's food is used to produce the heat, so availability of food in the environment could be construed to be part of the animal's physical environment. Food availability will not enter into our discussions in this way, but we will ask how much food a homeotherm needs in order to maintain constant body temperature. Such questions are of interest to those modeling ecosystems as well as those managing range lands for wild or domestic animals, because in many cases it is the only way an estimate of the food requirements for certain animals can be made.

The Energy Budget Concept

The question of whether or not an animal can maintain its body temperature within acceptable limits can be stated in another way which makes it more amenable to analysis. We need only ask whether heat loss can be balanced by heat input at the required body temperature. We are well prepared to describe heat inputs and heat losses for a system, so the problem is easily solved, at least in principle. An equation that says that the heat input minus the heat loss equals the heat storage for a system is called an energy budget equation. As an example of an energy budget, consider a representative unit area of the surface of an animal that is exposed to the atmosphere. The energy budget of this surface is the sum of the heat inputs and losses. Thus,

$$a_s S_i + a_L L_i - L_{oe} + M - \lambda E - H - q - G = 0 \qquad (7.1)$$

where a_s and a_L are the short- and long-wave absorptivities, S_i and L_i are the short- and long-wave irradiances at the surface, L_{oe} is the long-wave emittance, M is the metabolic heat per unit surface area supplied to the skin surface, H is the rate of convective heat loss, λE is the latent heat loss through evaporation of water from the respiratory tract and skin of the animal, G is the heat loss by conduction, and q is the rate of heat storage in the animal.

Initially, we will concern ourselves only with steady-state conditions for which the heat storage rate, q, is zero. Thus, we are interested in environments where heat losses and heat gains can be balanced. For simplicity, we will also assume $G = 0$. The $L_{oe}(=\epsilon\sigma T_s^4)$ and H $[=\rho c_p(T_s - T_a)/r_{Ha}]$ terms both involve the surface temperature of the animal. It should always be possible to find a value for surface temperature such that Equation 7.1 balances, but that temperature may be too high or too low for the animal to remain alive. If body temperature and metabolic rate are specified, then Equation 7.1 tells what environments are energetically acceptable. Thus we can specify food needs for survival in a given climate, or acceptable climates for a given food supply and activity level. To do this we need to look at each term in Equation 7.1 in detail.

Specifying Radiant Energy Fluxes for Animals

The first three terms in Equation 7.1 involve radiant energy exchange. The radiant flux densities are averages for the entire animal surface. Since different parts of the animal are exposed to different flux densities, we need to find a way to compute an average that will apply for the entire animal. We already know how to compute the long- and short-wave flux densities for various environments. These can be converted to the appropriate average values through multiplication by the ratio of the total body surface area of the animal to the surface area exposed to a particular radiant energy source.

Calculation is simplest for the long-wave emitted radiation (L_{oe}) since the emitting surface is the total surface area of the animal. The average emitted flux density is therefore just $L_{oe} = \epsilon\sigma T_s^4$, where ϵ is the average emissivity of the surface (same as long-wave absorptivity) and T_s is an appropriate average surface temperature. The incoming long-wave radiation is somewhat more difficult to find since the animal may be exposed to several sources of long-wave radiation, each having a different radiance. An appropriate average for a spherical absorber can be computed by multiplying the radiance of each source by the solid angle subtended by that source, and dividing the sum by 4π steradians. Few animals have spherical shapes, but this method can still be used to obtain approximate long-wave irradiances for real animals. A more detailed analysis using shape factors can be found in Kreith [7.4].

The simplest analysis is for an animal surrounded by a radiating surface at uniform temperature. In this case, $L_i = \epsilon_i \sigma T_i^4$ where T_i is the radiating wall temperature and ϵ_i is effective wall emissivity. If the enclosure is much larger than the animal, $\epsilon_i \rightarrow 1$ [7.3].

Another common situation is that in which the animal is outdoors. The upper hemisphere that surrounds him is radiating at sky temperature and emissivity, and the lower hemisphere is comprised of terrestrial objects radiating at ground temperature. If we assume the solid angle for each source to be about half of the total solid angle, the arithmetic average of the two flux densities will give the appropriate long-wave flux. Other more complicated arrangements can be handled in this same manner. Since the long-wave radiation is not directional, the total surface area of the animal receives the radiation.

The short-wave irradiance is composed of one directional component, the direct solar flux, and one nondirectional component, scattered and reflected short-wave radiation. The scattered and reflected short wave can be handled in the same way L_i was. One calculates the flux from each source and the solid angle subtended by the source. In the simplest case we need only consider short wave scattered from the sky and reflected from terrestrial objects. If the sky and objects make up about equal solid angles, we can average the contributions from each to obtain the nondirectional short-wave component.

The calculation of direct short-wave input for animals is probably most easily carried out by considering the projection of the animal on a plane perpendicular to the solar beam. The average direct short-wave irradiance is the ratio of projected area to total area multiplied by the irradiance perpendicular to the solar beam (S_p) (Equation 5.9). Clear sky S_p is relatively independent of elevation angle for most of the day. Thus a rough estimate for S_p on clear days would be 0.8 to 1 kW/m^2.

The projection of the animal on a surface perpendicular to the solar beam can be found by tracing the animal's shadow. It can also be calculated for simple geometric shapes such as spheres, cylinders, and spheroids. The simplest shape for this exercise is a sphere, since its projection is a circle. The ratio of projected area to total area is

$$\frac{A_p}{A} = \frac{\pi r^2}{4\pi r^2} = \frac{1}{4}$$

where r is the radius of the sphere. To find the average direct solar irradiance, we multiply S_p by A_p/A, so for a sphere, $S_{id} = 0.25\, S_p$. For other shapes, once we know A_p/A, calculation of S_{id} is accomplished in this same manner.

Figure 7.1 shows ratios of A_p/A for various other geometrical shapes that can be used to simulate animals. In each case, the angle θ is the angle between the solar beam and the axis of the body.

Finally, we need to know a_s and a_L to complete the specification of radiant energy absorption by the animal. Some typical values are given in Table 7.1. It should be kept in mind that the values for a_s apply only for solar radiation, and would be different for other sources.

Metabolism

Our energy budget equation requires metabolic rate in terms of energy supplied to the animal's surface per unit area. Physiologists generally measure metabolic rates per unit mass of animal. We therefore need to be able to convert from the commonly reported values to the ones we are interested in.

FIGURE 7.1. Ratios of shadow area on a surface perpendicular to the solar beam to total surface area for three shapes that can be used to simulate animal shapes. The angle θ is the angle between the solar beam and the axis of the solid.

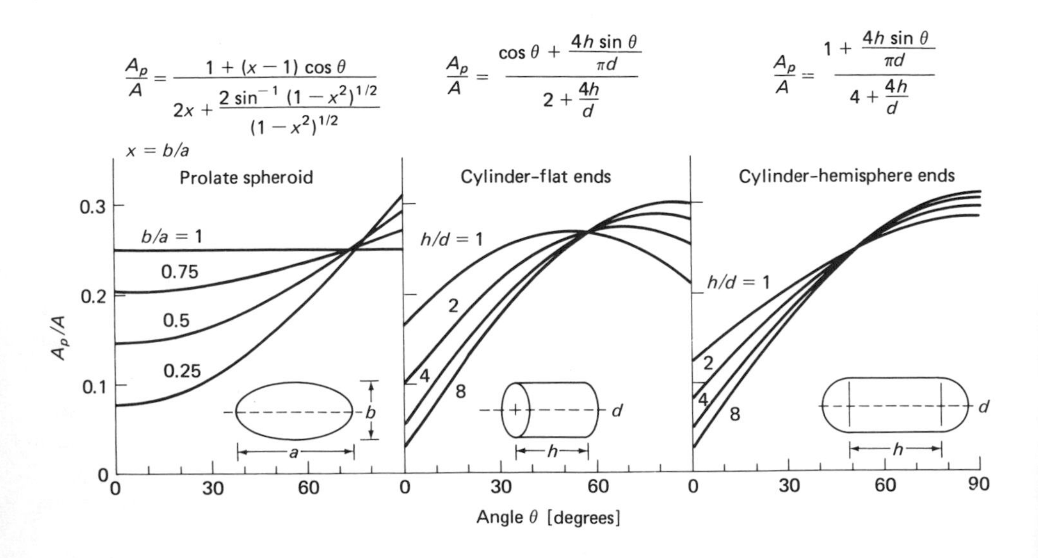

Table 7.1 Short-wave absorptivities for solar radiation and long-wave emissivities of animal coats

Mammals	Absorptivities		
	Dorsal	Ventral	Average
Red squirrel *(tamiasciurus hudsonicus)*	0.73	0.78	0.75
Gray squirrel *(Sciurus carolinensis)*	0.78	0.61	0.69
Field mouse *(Microtus pennsylvanicus)*	0.89	0.83	0.86
Shrew *(Sorex sp.)*	0.81	0.74	0.77
Mole *(Scalopus aquaticus)*	0.81	0.81	0.81
Gray fox *(Urocyon cinereoargenteus)*			0.66
Zulu cattle			0.49
Red Sussex cattle			0.83
Aberdeen angus cattle			0.89
Sheep, weathered fleece			0.74
Sheep, newly shorn			0.58
Birds	**Absorptivities**		
	Wing	Breast	Average
Cardinal *(Cardinalis cardinalis)*	0.77	0.60	
Bluebird *(Sialia sp.)*	0.73	0.66	
Tree swallow *(Iridoprocne bicolor)*	0.76	0.43	
Magpie *(Pica pica)*	0.81	0.54	
Canada goose *(Branta canadensis)*	0.85	0.65	
Mallard duck *(Anas platyrhynchos)*	0.76	0.64	
Mourning dove *(Zenaida macroura)*		0.61	
Starling *(Sturnus vulgaris)*			0.66
Glaucous-winged gull *(Larus glaucescens)*			0.48
Animal	**Emissivities**		
	Dorsal	Ventral	Average
Red squirrel *(Tamiasciurus hudsonicus)*	0.95–0.98	0.97–1.0	
Gray squirrel *(Sciurus carolinensis)*	0.99	0.99	
Mole *(Scalopus aquaticus)*	0.97		
Deer mouse *(Peromyscus sp.)*		0.94	
Gray wolf *(Canis lupus)*			0.99
Caribou *(Rangifer arcticus)*			1.00
Snowshoe hare *(Lepus americanus)*			0.99

Sources of data are given by Monteith [7.7].

Control of body temperature for homeotherms and maintenance of body functions in all animals requires a minimal or basal metabolic rate. This basal rate (Watts) can be approximated for a wide variety of animals by the equation

$$B_m = Cm^{3/4} \tag{7.2}$$

where m is the animal's mass (kg) and C is a constant, generally taken as 3 to 5 for homeotherms and around 5 percent of this value for poikilotherms at 20°C. Figure 7.2 shows this relationship for a wide range of animal sizes. An approximate relation between surface area (m^2) and mass (kg) is

$$A = 0.1\, m^{2/3}. \tag{7.3}$$

The uncertainty in the exponents of Equations 7.2 and 7.3 is large enough that, for many practical purposes, they may be taken as being the same. Thus the basal metabolic rate per unit animal surface area, $M_b(=B_m/A)$, is relatively independent of animal size. Typical values for M_b range from 30 to 50 W/m^2.

Metabolic rate increases with the animal's activity. We can account for this in the energy budget in one of two ways. If we assume about 30-percent efficiency for conversion of chemical energy to work in animals, then for each unit of work done there will be about two units of heat produced. If we know how much work is done we can calculate the metabolic production of heat. The other method is somewhat simpler. As a rule of thumb, we can assume that the maximum aerobic metabolic rate an animal can sustain is about 10 times the basal rate (Figure 7.2). If we can estimate the animal's activity as a percent of maximum (say from oxygen consumption measurements or running speed compared to maximum) we can estimate the metabolic contribution from

FIGURE 7.2. Relation between body mass and basal metabolic rate of poikilotherms at 20°C, basal metabolic rate of homeotherms at 39°C, and maximum aerobic metabolic rate.

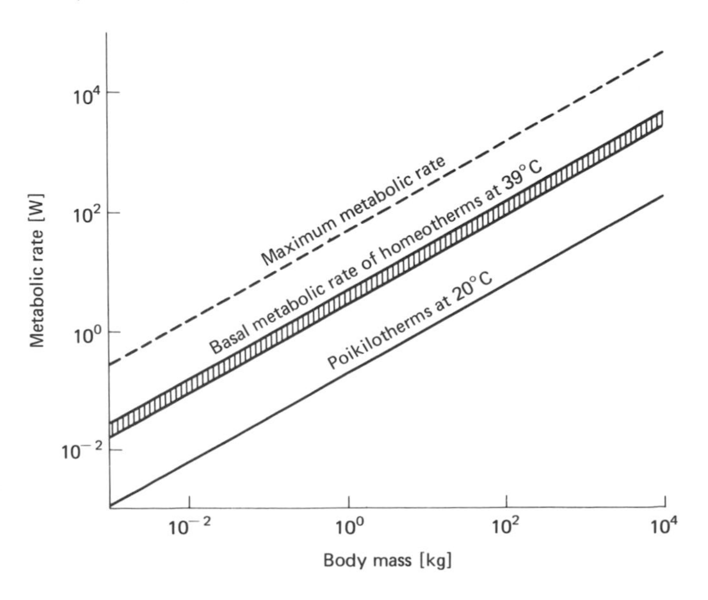

$$M = M_b\left(1 + \frac{9\,\alpha}{\alpha_M}\right) \tag{7.4}$$

where α is the animal's activity and α_M is the maximum sustainable activity. If M_b is 50 W/m², M will vary between 50 and 500 W/m².

Latent Heat Exchange

Evaporation of water from the respiratory tract and from the skin result in latent heat loss from the animal. In respiratory evaporation, air is breathed in at atmospheric vapor density and breathed out at the saturation vapor density at body temperature (except for some animals with special water-conserving mechanisms which we will discuss later). A reasonable assumption would be that respiratory latent heat loss is some fixed fraction of metabolic heat production since increased metabolic heat production results in increased oxygen consumption and this increases breathing rate. The ratio of respiratory evaporation to metabolic production can be calculated from

$$\frac{\lambda E_R}{M} = \frac{(\rho_{ve} - \rho_{vi})\,\lambda}{(\rho_{oi} - \rho_{oe})\,\Gamma} \tag{7.5}$$

where ρ_v is vapor density in air, ρ_o is oxygen density in air, i is the inspired concentration, e is the expired concentration, λ is the latent heat of vaporization, and Γ is the heat produced per kilogram of oxygen consumed. If we take $\lambda = 2.4$ MJ/kg, $\Gamma = 15$ MJ/kg, $\rho_{ve} = 44$ g/m³, $\rho_{vi} = 7$ g/m³, $\rho_{oi} = 280$ g/m³, and $\rho_{oe} = 210$ g/m³ (corresponding to a reduction in oxygen concentration from 21 to 16 percent by volume), Equation 7.5 gives $\lambda E_R/M = 0.08$.

Some animals with small nasal passages exhale air at temperatures well below body temperature. Figure 7.3 compares exhaled air temperature for several bird species with values for man and for kangaroo rats *(Dipodomys merriami)*. The exhaled air temperature for the rat is even lower than air temperature, and approaches wet bulb temperature. The appropriate value for ρ_{ve} for these animals would be the saturation vapor density at exhaled air temperature rather than body temperature. The effect of cooling the air is to conserve body water. As an example of this, we can find the ratio of $\lambda E_R/M$ for a kangaroo rat at 20°C. From Figure 7.3, $T_{\text{expired}} = 18$°C and $\rho_{ve} = 15$ g/m³. If the other values are as in our previous example, $\lambda E_R/M = 0.02$. This is only 25 percent of the respiratory water loss per unit area of a man under similar conditions. This is important for survival of kangaroo rats in their arid habitat.

Cutaneous water loss from animals can be calculated from

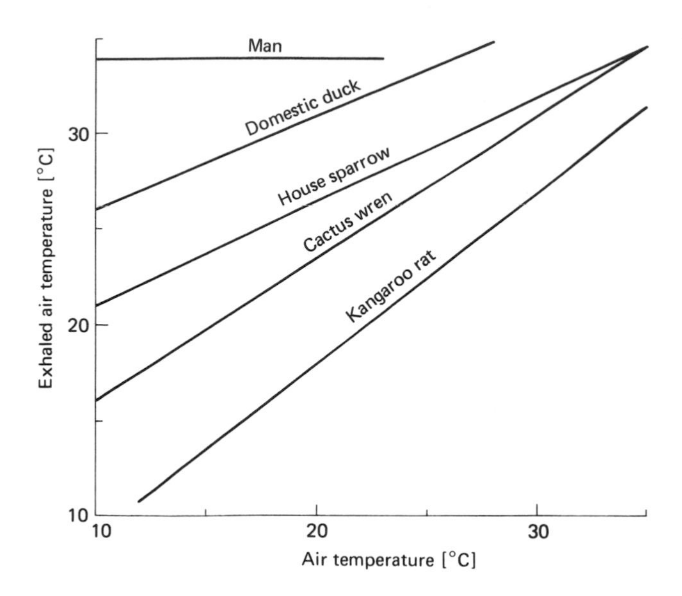

FIGURE 7.3. Temperature of exhaled air as a function of air temperature for several species. (After Schmidt-Nielsen [7.12].)

$$E_s = \frac{\rho'_{vs} - \rho_{va}}{r_{vs} + r_{vc} + r_{va}} \tag{7.6}$$

where r_{vs}, r_{vc}, and r_{va} are the resistances to vapor diffusion through the skin, coat, and boundary layer, and ρ'_{vs} is the saturation vapor density at skin surface temperature. For animals with moist skins (earthworms, snails, amphibians) $r_{vs} = r_{vc} = 0$, and the controlling resistance for water loss is the boundary layer resistance. For nonsweating animals, r_{vs} is often so large that r_{vc} and r_{va} are negligible by comparison. Table 7.2 shows skin resistances to vapor diffusion for nonsweating animals in dry air. Note that species that survive in dry environments tend to have the highest vapor-diffusion resistances. Little is known about the variability of these numbers or their dependence on environmental moisture or temperature. Much additional research is needed in this area. Accurate estimates of skin-diffusion resistance are important, both for accurate energy budget predictions and for water budgets of animals. The importance of skin water loss is illustrated by the fact that it accounts for 75 percent or more of the total water loss even for the desert tortoise [7.11].

To illustrate the magnitude of λE_s, we will find the rate of skin water loss for a camel under circumstances similar to those for which we found λE_R. If we assume, as before, $\rho_{va} = 7$ g/m^3, $\rho'_{vs} = 44$ g/m^3, and $r_{vs} = 13000$ s/m, then

Table 7.2 Skin resistance to vapor diffusion for animals (not heat stressed)

Animal	r_{vs} (ks/m)	Reference for Data
Mammals		
White rat *(Rattus sp.)*	3.9	[7.11]
Man *(Homo sapiens)*	7.7	[7.3]
Camel *(Camelus sp.)*	13	[7.11]
White footed mouse *(Peromyscus sp.)*	14	[7.11]
Spiny mouse *(Acomys sp.)*	15	[7.11]
Reptiles		
Caiman *(Caiman sp.)*	5.5	[7.11]
Water snake *(Natrix sp.)*	8.8	[7.11]
Pond turtle *(Pseudemys sp.)*	15	[7.11]
Box turtle *(Terrapene sp.)*	33	[7.11]
Iguana *(Iguana sp.)*	36	[7.11]
Gopher snake *(Pituophis sp.)*	40	[7.11]
Chuckawalla *(Sauromalus sp.)*	120	[7.11]
Desert tortoise *(Gopherus sp.)*	120	[7.11]
Birds		
Sparrow *(Zonotrichia leucophrys gambelii)*	7.6	[7.10]
Budgerigar *(Melopsittacus indulatus)*	8.5	[7.1]
Zebra finch *(Poephila guttata)*	10.2	[7.1]
Village weaver *(Ploceus cucullatus)*	12.4	[7.1]
Poor-will *(Phalaenoptilus nuttallii)*	13.2	[7.5]
Roadrunner *(Geococcys californianus)*	17.2	[7.5]
Painted quail *(Excalfactoria chinensis)*	20	[7.1]
Ostrich *(Struthio camelus)*	56	[7.13]

$$\lambda E_s = \frac{2430 \text{ J g}^{-1}(44 - 7) \text{ g m}^{-3}}{13000 \text{ s m}^{-1}} = 6.9 \text{ W/m}^2.$$

If we assume $\lambda E_R/M = 0.08$ and $M = 50$ W/m², then $\lambda E_R = 4$ W/m². The skin latent heat loss is about 63 percent of the total, and the total latent heat loss is around 20 percent of M. These percentages are probably typical of homeotherms that are not heat-stressed. For poikilotherms under similar conditions it is often possible to assume $M = \lambda E$ without introducing appreciable error in the energy budget equation.

As the animal becomes heat-stressed, latent heat loss increases, generally by some active process such as sweating or panting. There is no general approach to calculation of latent heat loss under these conditions since animal responses are so varied. The approach would need to be fitted to the particular species being studied. In Chapter 8 we will look at latent heat loss by sweating, but will not otherwise treat water loss under heat stress.

Conduction of Heat in Animal Coats and Tissue

Steady-state heat conduction in animals is described by Equation 6.1 with the appropriate expression for resistance chosen from Equations 6.6 to 6.9, depending on the geometry of the system. Where heat is conducted through several layers having different thermal resistances, the thermal resistance of each layer can be determined and the sum of the component resistances used as the total resistance to heat transfer.

Our concern, then, needs to be with finding appropriate values of D_H for animal coats and tissue. Generally handbooks list the thermal conductivity (k) of materials. Conductivities can be converted to apparent diffusivities appropriate for use in our equations using $D_H = k/\rho c_p$, where ρ and c_p are the density and specific heat of air. This conversion has no particular physical significance; ρc_p is used, as in Equation 6.1, to convert k to appropriate units. The advantage of using the values of ρc_p for air in all calculations rather than determining the actual diffusivity is that resistances can then be added to or compared directly with boundary layer resistance.

As an example, we will calculate the thermal resistance of a 1-cm-thick layer of cotton. A handbook value for k is 0.059 W $m^{-1}K^{-1}$. Division by ρc_p gives $D_H = 0.059$ J $m^{-1}K^{-1}s^{-1}$/1200 J $m^{-3}K^{-1} = 4.9 \times 10^{-5}$ m^2/s. The thermal resistance of a 1-cm-thick plane layer (Equation 6.7) is $r_H = l/D_H = 10^{-2}$ m/4.9×10^{-5} m^2s^{-1} = 204 s/m. Using D_H for air from Table A.1 we can calculate the resistance of a 1-cm-thick layer of still air. It is $r_H = 10^{-2}/2.15 \times 10^{-5} = 465$ s/m, so the cotton is only 44 percent as effective as still air in reducing heat loss. Cena and Montieth [7.2] have shown that free convection and radiaiton, as well as conduction through still air trapped by the coat, contribute significantly to heat transfer in animal coats. They found ''radiative conductivities'' ranging from 0.005 to 0.02 W $m^{-1}K^{-1}$ depending on coat characteristics (density and color). Conductivities for still air, radiation, and free convection are added to give the total coat conductivity. The coat resistance is then determined as was shown in the example. The thermal conductivity of still air is 0.026 W $m^{-1}K^{-1}$. If for an animal with a 4-cm-thick coat and negligible free convection the radiative conductivity were 0.02 W $m^{-1}K^{-1}$, the coat thermal conductivity would be $0.026 + 0.02 = 0.046$ W $m^{-1}K^{-1}$. The thermal resistance of the coat would be 1043 s/m, a little over half the resistance of an equivalent depth of still air.

Figure 7.4 gives thermal resistances of small patches of animal skin for several species. These measurements were made under laboratory conditions. We would expect the values obtained to be highly dependent on the conditions under which

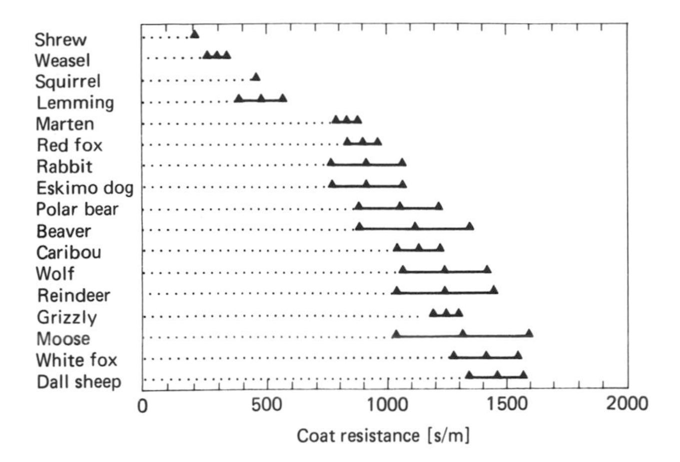

FIGURE 7.4. Thermal resistance of animal coats measured on pieces of raw skin and fur. (After Scholander *et al.* [7.14].)

the measurements were made, so some caution must be exercised in using these values. Also, in dealing with real animals, coat depth varies from point to point, and we need an average thermal resistance for the entire body. For this reason, thermal resistances determined on live animals [7.10] may be much more useful than those estimated on portions of animal coats. However, in the absence of whole animal data, the foregoing arguments and Figure 7.4 can be used to estimate thermal resistance of coats. The unit ''clo'' has been used in numerous human and animal energy budget studies. It is defined as the thermal resistance required to keep a resting man comfortable in a 21 °C room. Such a definition lacks the precision necessary for scientific studies, and we will not use it in our analyses. The numerical value given for 1 clo is equivalent to $r_H = 200$ s/m, or about the same value as 1-cm thickness of cotton.

Wind, of course, can have a major effect on the thermal resistance of clothing and animal coats. The few studies that have been done tend to indicate that average coat resistance decreases with the square root of windspeed. Thus one could write

$$r_{Hc} = r_{Hc}^0 - b\sqrt{u}$$

where r_{Hc}^0 is the resistance with no wind and b is the decrease per unit change in $\sqrt{u}$. Studies on sparrows and rabbit pelts both gave b values near 50 [7.7, 7.10]. The value of b will depend on the penetration of the coat by wind and on the effect of wind on coat thickness, so one would expect considerable variation among species.

Heat transfer to the skin surface of an animal depends on blood flow, and is subject to regulation, within limits, by vasoconstriction or vasodilation. This regulation is important in control of body temperature. Table 7.3 shows maximum and minimum values of average tissue resistance for several species. As with coat resistance, these values are for a specific set of conditions. The maxima and minima can be altered dramatically through acclimatization. However, they can be used to estimate thermal resistance in the absence of other data.

Application of Energy Budget Equations

Equation 7.1, though useful, contains one term that is difficult to measure or estimate, that being the average surface temperature. We can make use of the tissue and coat resistance information available to us to eliminate T_s from the energy budget equation. Heat transfer from the animal's core to the exchange surfaces can be expressed as

$$M - \lambda E = \frac{\rho c_p (T_b - T_s)}{r_{Hb}} \tag{7.7}$$

where T_b is the animal's deep body temperature, T_s is the surface temperature, and r_{Hb} is the average coat-plus-tissue heat transfer resistance. We are assuming that all latent heat loss is from inside the animal. Our previous discussion indicated that this is a poor assumption, but it can be shown that the error from it is negligible [7.10] for nonsweating animals. The average surface temperature of the animal can now be written as

$$T_s = T_b - \frac{r_{Hb}(M - \lambda E)}{\rho c_p}. \tag{7.8}$$

If we write Equation 7.1 explicitly (assuming $q = G = 0$), it becomes

Table 7.3 Thermal resistance of peripheral tissue of animals

Animal	Vasoconstriction resistance (s/m)	Vasodilation resistance (s/m)
Steer	170	50
Calf	110	50
Pig (3 months)	100	60
Down sheep	90	30
Man	90	15

Data from Kerslake [7.3] and Monteith [7.7].

$$a_s \frac{A_p}{A} S_p + a_s \overline{S}_d + a_L \overline{L}_i - \epsilon\sigma T_s^4 + M - \lambda E - \frac{\rho c_p (T_s - T_a)}{r_{Ha}} = 0. \quad (7.9)$$

S_p is the solar irradiance perpendicular to the the beam, $\overline{S}_d$ is the mean diffuse solar irradiance (scattered and reflected), and $\overline{L}_i$ is the average incoming long-wave irradiance (sky and ground). A Taylor expansion of Equation 7.8, with the higher order terms neglected, is substituted for T_s^4 in Equation 7.9, and Equation 7.8 is substituted directly for T_s in the convection term. With these substitutions, Equation 7.9 becomes

$$R_{\text{abs}} - \epsilon\sigma T_b^4 + M - \lambda E - \frac{\rho c_p (T_b - T_a)}{r_{Ha}} + \frac{r_{Hb}(M - \lambda E)}{r_e} = 0. \quad (7.10)$$

where

$$R_{\text{abs}} = a_s \frac{A_p}{A} S_p + a_s \overline{S}_d + a_L \overline{L}_i, \qquad \frac{1}{r_e} = \frac{1}{r_r} + \frac{1}{r_{Ha}}$$

and $r_r = \rho c_p / 4\epsilon\sigma T^3$ is a "resistance" to long-wave radiative transfer. Equation 7.10 is the animal energy budget equation in a form that is easily cvaluated. Note that the first five terms describe heat exchange for an animal with $T_s = T_b$ or $r_{Hb} = 0$. The last term is a correction applied to the "no-resistance" animal to give the heat exchange for a real animal.

The Climate Space

Equation 7.10 can be graphically displayed in such a way as to indicate energetically acceptable environments for animals. Such a graph is called a *climate diagram* and defines a *climate space* for the animal [7.9]. Equation 7.10 contains terms related to environment (absorbed radiation, wind and air temperature) and terms related to the animal (metabolic rate, body heat transfer resistance, and body temperature). For a given metabolic rate, body temperature, and tissue-plus-coat resistance, Equation 7.10 describes a straight-line relationship between air temperature and absorbed radiation. The slope of this line is $\rho c_p / r_{Ha}$ and the intercept is determined by T_b, r_{Hb}, and $M - \lambda E$. Figure 7.5 shows this line for two metabolic rates and two windspeeds for a masked shrew *(Sorex cinereus)*. The metabolic rates are 50 and 150 W/m², which probably represent the minimum and maximum values that are sustained for long time periods. The latent heat loss is assumed to equal 0.2 M.

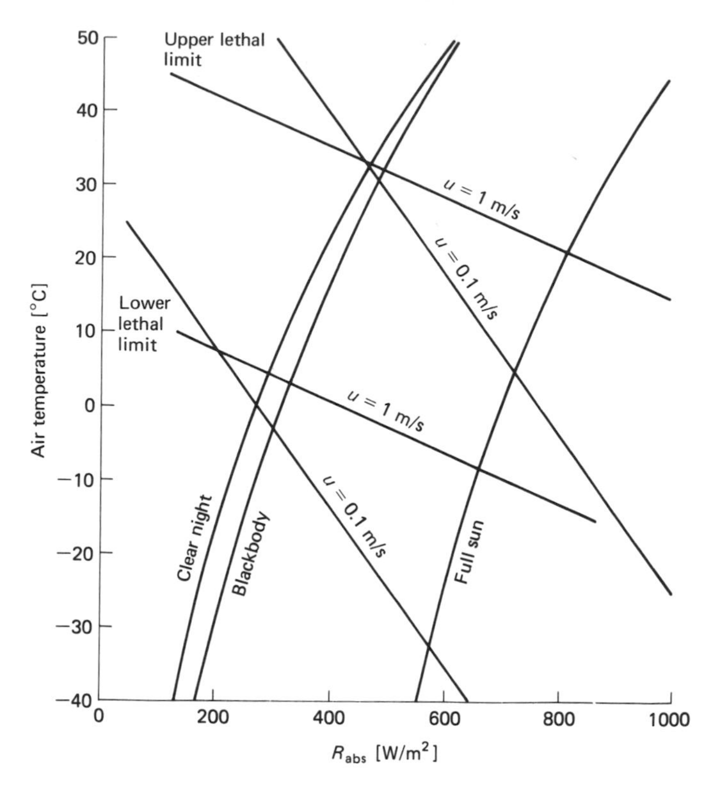

FIGURE 7.5. Climate diagram for a masked shrew. See text for assumptions.

The two windspeeds chosen are 0.1 and 1 m/s, which probably represent extremes for the shrew's environment. The deep body temperatures chosen for the two lines are 41 and 37.5°C. Again these are probably limits for survival of the shrew. The body resistance (r_{Hb}) for the shrew is taken as 300 s/m (Figure 7.4 and Table 7.3) at $T_b = 37.5$ and 230 s/m at $T_b = 41$. These numbers were chosen to represent extremes for survival. Thus the upper lines represent the upper lethal limit (U.L.L.) for survival of the shrew and the lower lines are the lower lethal limit (L.L.L.).

Three other lines make the climate space more useful. These describe physically attainable environments for the shrew. The blackbody line represents the radiation-temperature relationship for a blackbody cavity (such as a burrow) where wall temperature equals T_a. The equation for this line is just $R_{\text{abs}} = \sigma T_a^4$. The clear night sky line is the minimum radiant energy environment available to the shrew. It is described by the equation

$$\text{Night } R_{\text{abs}} = a_L \bar{\epsilon}_s \sigma T_a^4 \tag{7.11}$$

where $\bar{\epsilon}_s$ is the average emissivity of the surroundings (clear sky and ground).

The bright sun line represents the maximum energy available to the animal under a clear sky. It is calculated from

$$\text{Sun } R_{\text{abs}} = \text{Night } R_{\text{abs}} + a_s S_p \frac{A_p}{A} + a_s \bar{S}_d. \tag{7.12}$$

From Table 7.1, a_s for shrew = 0.77. If we take S_p = 1000 W/m^2, $\bar{S}_d$ = 250 W/m^2, and A_p/A (Figure 7.1 with d = 1.8 cm, h = 3.6 cm, θ = 90) = 0.3, then Sun R_{abs} = Night R_{abs} + 424 W/m^2. We see that Night R_{abs} is relatively unaffected by animal characteristics, but Sun R_{abs} depends quite heavily on them.

We note now (Figure 7.5) that the Sun R_{abs}, Night R_{abs}, U.L.L., and L.L.L. lines enclose a space on the climate diagram. Combinations of radiation and temperature that fall within the space are energetically acceptable and environmentally possible for the animal. The animal could not survive if it were continuously exposed to environments outside the space. Field observation of animal behavior should indicate an overwhelming preference for environments inside the climate space, and, in addition, a marked preference for environments toward the center of the space. It is interesting to note (Figure 7.5) that the shrew seems better able to withstand cold than heat. In full sun and low wind, the highest air temperature he could withstand would be 5°C.

The climate space for a poikilotherm looks somewhat similar to that of a homeotherm, but the strategy is quite different. Instead of changing resistance and metabolic rate, the poikilotherm changes body temperature. Figure 7.6 shows a climate space for the desert iguana *(Dipsosaurus dorsalis)*. As a first approximation, we will assume $M - \lambda E = 0$ for the iguana. This results in considerable simplification of the energy budget equation. Lines for four body temperatures are shown. The highest and lowest lines represent U.L.L. at T_b = 45°C and L.L.L. at T_b = 3°C. Lines at T_b = 40°C and 30°C, represent the normal operating temperature range of the iguana. Windspeed is 0.3 m/s, and d is taken as 1.5 cm. Two values of a_s are used; a_s = 0.6 for light-colored animals and a_s= 0.8 for dark. For A_p/A = 0.3, Sun R_{abs} = Night R_{abs} + 550 a_s. Note that the iguana is much better suited to hot environments than the shrew, but not well suited to cold environments. The higher a_s of the dark animal gives it some advantage in calm, cool, sunny environments, but some disadvantage in hot, calm, sunny environments. At higher windspeeds, the effect is small. Actually, from a survival standpoint, color-

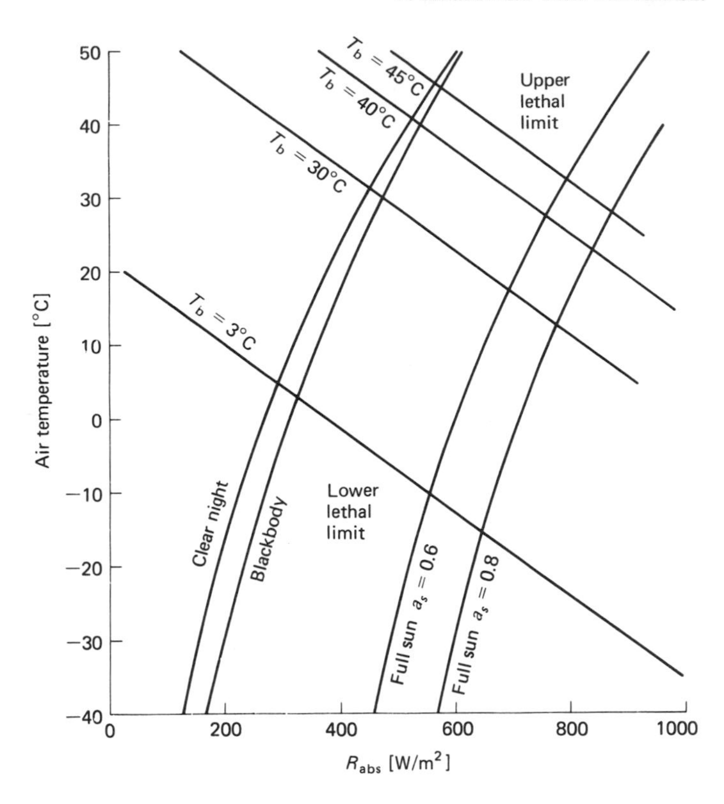

FIGURE 7.6. Climate diagram for desert iguana for a wind speed of 0.3 m/s. See text for assumptions.

ing may be more important as a protection from predators than in significantly altering the energy budget.

The Equivalent Blackbody Temperature

The climate diagram provides a convenient way for us to combine environmental variables. It allows us to visualize the equivalence of a high-temperature environment having little radiant energy input and a low-temperature environment having high radiant energy. We can accomplish this same thing mathematically. We will look for the temperature of a blackbody cavity (having wall temperature equal to air temperature) that will provide the same radiant and convective exchange as is present in the animal's natural environment. We will call this temperature the equivalent blackbody temperature [7.8] with symbol T_e. In Figures 7.5 and 7.6, T_e is the temperature at the intersection of the animal energy budget line (sloping downward to the right) and the blackbody line. Thus T_e combines the two environmental variables, T_a and R_{abs} into one variable. The energy budget equation for an animal in a blackbody cavity (Equation 7.10) is

$$\epsilon\sigma T_e^4 - \epsilon\sigma T_b^4 + M - \lambda E - \frac{\rho c_p(T_b - T_e)}{r_{Ha}}$$

$$+ \frac{r_{Hb}(M - \lambda E)}{r_e} = 0. \tag{7.13}$$

If we equate Equation 7.13 with 7.10 and solve for R_{abs} we find

$$R_{abs} = \frac{\rho c_p(T_e - T_a)}{r_{Ha}} + \epsilon\sigma T_e^4. \tag{7.14}$$

Using Equation 7.14 as a definition of T_e, we obtain a very convenient form of the energy budget equation for animals:

$$\frac{\rho c_p(T_b - T_e)}{r_{Hb} + r_e} = M - \lambda E. \tag{7.15}$$

This equation simply describes sensible and radiant heat loss from an animal in a blackbody cavity with temperature T_e. The total heat exchange resistance is the sum of r_{Hb} and the equivalent parallel resistance of r_{Ha} and r_r (Equation 7.10). For a poikilotherm, we have assumed $M - \lambda E = 0$, so Equation 7.15 predicts $T_b = T_e$. A little study of Equation 7.15 reveals that T_e will give us all of the information contained in the climate diagram analysis. If we supply extreme values for T_b, r_{Hb}, r_e, M, and λE, then Equation 7.15 will show what values of T_e are energetically acceptable. In addition, if we supply values for T_b, T_e, r_{Hb}, r_e, and λE, we can predict how much food energy the animal will need for thermoregulation. Apparently measurement and prediction of T_e is quite important to animal energy budget studies.

As we already pointed out, when $M - \lambda E = 0$, $T_b = T_e$. Thus, one way to find T_e would be to use a model animal having convective and radiant exchange properties similar to the real animal. By placing the model in the animal's environment and measuring T_b of the model, one would know T_e.

Another method for finding T_e would be to solve for it from Equation 7.14, however this is complicated by the fact that the equation is a quartic in T_e. An approximate solution can be obtained by writing

$$T_e = T_a + \Delta T$$

with

$$\Delta T = T_e - T_a$$

then

$$T_e^4 = (T_a + \Delta T)^4$$

$$= T_a^4 + 4T_a^3\Delta T + 6T_a^2(\Delta T)^2 + 4T_a(\Delta T)^3 + (\Delta T)^4.$$

This same type of expansion was used to obtain Equation 7.10 from Equations 7.8 and 7.9. Since ΔT is small compared to T_a, squared, cubed, and fourth-power terms in ΔT will be much smaller than the first two terms of the expansion and

$$T_e^4 \simeq T_a^4 + 4T_a^3(T_e - T_a). \tag{7.16}$$

This approximation is good enough as it stands for most purposes, but can be improved by using $\overline{T}^3$ in the second term rather than T_a^3, where $\overline{T}$ is the average of T_a and T_e. Equation 7.16 can be substituted into Equation 7.14 to give

$$T_e = T_a + \frac{r_e(R_{abs} - \epsilon\sigma T_a{}^4)}{\rho c_p} \tag{7.17}$$

with r_e defined, as in Equation 7.10, as the equivalent parallel resistance of r_{Ha} and r_r. Note that when $R_{abs} = \epsilon\sigma T_a^4$, $T_e = T_a$. This condition might apply for animals in burrows, under dense crop or cloud cover, or in experimental enclosures. When the animal is exposed to solar radiation or to a clear sky, the second term in Equation 7.17 can be thought of as a radiant energy increment or decrement to air temperature. As an example, we can find the room temperature (T_e) which is equivalent to an outdoor temperature of 30°C for a human in full sun and a 1 m/s wind. If we assume $\theta = 60°$, then $A_p/A = 0.26$ (Figure 7.1). For dark clothing, we could take $a_s = 0.8$ and $\epsilon \simeq 1$, so

$$R_{abs} = a_s\left(\frac{A_p}{A} S_p + \overline{S}_d\right) + \overline{\epsilon}_s \sigma T_a^4$$

$$= 0.8(0.26 \times 1000 + 250) + 0.94 \times 479$$

$$= 858 \text{ W/m}^2$$

A characteristic dimension frequently used for humans is $d = 0.17$ m, so

$$r_{Ha} = 0.7 \times 307\left(\frac{d}{u}\right)^{1/2} = 215\left(\frac{0.17}{1}\right)^{1/2} = 89 \text{ s/m}$$

$$r_r = \frac{\rho c_p}{4\epsilon\sigma T_a^3} = \frac{1200}{6.3} = 190 \text{ s/m}$$

$$r_e = \frac{r_{Ha} r_r}{r_{Ha} + r_r} = \frac{89 \times 190}{89 + 190} = 60 \text{ s/m}$$

$$T_e = 30 + \frac{60\,(858 - 479)}{1200} = 49°\text{C}.$$

If one found himself in this environment, he would probably be looking for some shade. If windspeed were increased to 3 m/s, T_e would decrease to 43°C. If white clothing with a_s = 0.3 were worn, T_e would be 36°C at u = 1 m/s.

Besides being useful for predicting animal behavior (as in Figures 7.5 and 7.6) energy budget equations can be used to predict food requirements. As an example, we could find the food requirement for a 1.5-kg rabbit in an environment with T_e = 0°C. When averaged over day and night conditions, T_e is probably about equal to T_a for a typical rabbit environment. For our calculations we can assume u = 1 m/s and d = 10 cm, though we will see that the exact values used have little effect on the result. We find that r_{Ha} = 68 s/m, r_r = 261 s/m, and r_e = 54 s/m. From Figure 7.4, r_{Hb} = 1000 s/m for rabbit. From Equation 7.15

$$M - 0.2M = \frac{1200\,(37 - 0)}{1000 + 54}$$

or

$$M = 53 \text{ W/m}^2.$$

We have assumed $\lambda E/M$ = 0.2 and T_b = 37°C. From Equation 7.3, the area of a 1.5-kg animal is around 0.13 m, so the rabbit's energy requirement is 0.13 × 53 = 6.9 W. If we take the caloric content of glucose as 15.7 MJ/kg, a kilogram of glucose would last 15.7 × 10^6/6.8 = 2.3 × 10^6 s, or 27 days. One kilogram of dry grass might last only half that long because of inefficiencies in absorption in the digestion process. Efficiency factors are known for many animals and diets. One thing to keep in mind is that this is only the energy cost of thermoregulation. Though this is generally the primary energy cost for homeotherms, the total food requirement will be higher than our calculations show.

The Transient State

Obviously, animals are not continuously in environments that fall within their climate space. Short periods of intense activity, such as running, flying, or climbing, or exposure to high winds at cold temperature are common. During these times the q term in Equation 7.1 is not zero, and there is a positive or negative storage of heat in the body. Transient state energy budgets are important to the animal, but have not been examined in much detail by researchers.

If we can assume that the termal conductivity and heat capacity of an animal's core are large compared to that of the coat and peripheral tissues, we can write

$$q = \frac{V}{A} \rho_b c_b \frac{dT_b}{dt} \tag{7.18}$$

where V is the body volume and ρ_b and c_b are the body density and specific heat. Equation 7.15 can be rewritten to include transient effects as

$$M - \lambda E - \frac{\rho c_p (T_b - T_e)}{r_{Hb} + r_e} = \frac{V}{A} \rho_b c_b \frac{dT_b}{dt}. \tag{7.19}$$

Equation 7.19 is not very useful for homeotherms at temperatures within their control band because M, λE, and r_{Hb} are functions of T_b or some other temperature generally related to T_b. Integration is therefore difficult, if not meaningless. If, for poikilotherms, we assume $M - \lambda E = 0$, then integration of Equation 7.19, taking all but T_b as constant, gives

$$\frac{T_{b2} - T_e}{T_{b1} - T_e} = e^{-t/\tau} \tag{7.20}$$

with

$$\tau = \frac{\rho_b c_b V (r_{Hb} + r_e)}{\rho c_p A}.$$

Equation 7.20 can be used to find the time required for a poikilotherm to change from T_{b1} to T_{b2} when T_e is changed. The symbol τ represents the "time constant" of the animal. It has units of seconds, and is an index of response time. At $t = \tau$ the system will have changed to within $1/e = 0.37$ of the total change from T_{b1}, the initial temperature, to T_e, the final temperature. Thus animals with large time constants could survive exposure to environments outside their climate space for relatively long times. From Equation 7.20 we note that large volume, small surface area, and large thermal resistance maximize τ. Thus animals that weigh only a few grams would respond quickly to environmental changes and would always be close to equilibrium conditions. Animals weighing many kilograms would seldom be at steady state and could survive short-term extremes in exposure much more readily.

A transient analysis could also be used to indicate thermal behavior of homeotherms when they are subjected to environments that are sufficiently harsh to force complete commitment of their temperature control systems. Under these conditions M, λE, and r_{Hb} become constant at their maximum or minimum values, and it is possible to integrate Equation 7.19. The resulting equation is similar to Equation 7.20 except that the final temperature, rather than being T_e, is $T_e + (M - \lambda E)(r_{Hb} + r_e)/\rho c_p$ because of metabolic heat and evaporation.

Animals and Water

The water budget for an animal can be written in the same form as the energy budget, namely: Water in − water out = stored water. Unlike the energy budget, the water budget can seldom be analyzed as a collection of steady-state processes. Intake is from free-water sources, metabolism, and water in the animal's food. Water is taken in discrete events, not continuously. As we have seen, cutaneous and respiratory water loss are relatively continuous, and determined largely by animal activity and environment. Water is also lost in feces and urine, though this is often only a small fraction of the total water loss. Water loss reduces the amount of water stored in the blood, tissues, or digestive tract of the animal, and may decrease the osmotic potential (increase concentration) of body fluids.

MacMillen [7.6] had used urinary osmotic potential of kangaroo rats *(Dipodomys merriami)* in the desert as an index of water balance. His data (Figure 7.7) show only slight variations in plasma osmotic potential through the season. Urine osmotic potential decreased in the summer and increased in the winter, indicating higher water deficits in the summer. Urine osmotic potentials of some other desert rodent species were less well correlated with seasonal temperature changes because of changes in diet with season. For comparison, the osmotic potential of human plasma is −0.75 kJ/kg and human urine is generally between −2.1 and −3.3 kJ/kg.

MacMillen also conducted laboratory studies on several desert rodent species to determine their ability to maintain water balance on a diet of dry birdseed. The kangaroo rat neither lost nor gained weight, but some other species fell far short of maintaining positive water balance. Others actually gained weight. The pocket mouse *(Perognathus longimembris)* seemed to be able to maintain a particularly favorable water balance on this diet. One wonders first how these ani-

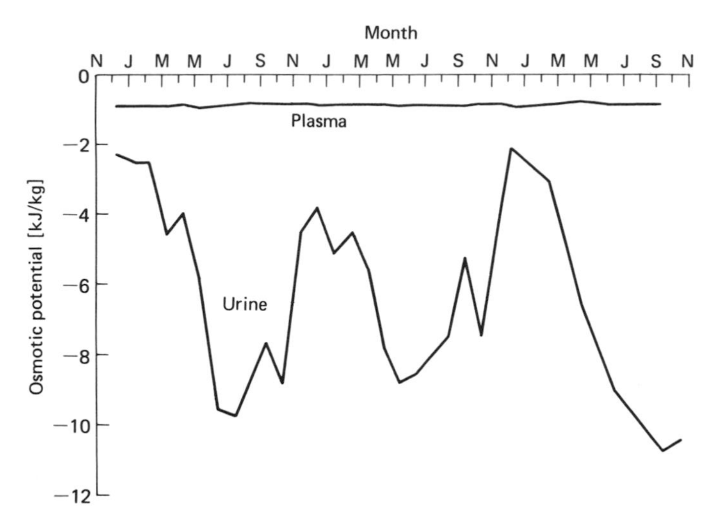

FIGURE 7.7. Osmotic potentials of plasma and urine of desert dwelling *Dipodomys merriami* over a three-year period. (After MacMillen and Christopher [7.6].)

mals can get enough water from dry seeds to supply their needs, and second why the kangaroo rat would have a less favorable water balance than the smaller mouse. Some light can be shed on these questions by a simple analysis.

When an animal oxidizes food to produce heat, water is also produced. One kilogram of glucose, when oxidized, produces 600 g of water. The ratio of latent heat from respiratory water to metabolic heat produced is $\lambda E_{pr}/M = 0.1$. We have already found the respiratory latent heat loss for the kangaroo rat at 20°C to be 0.02 M. The skin latent heat loss (Equation 7.6) is 5 W/m², if we assume $\rho'_{va} - \rho_{va} = 30$ g/m³ and $r_{vs} = 15$ ks/m (Table 7.2). Equation 7.15 can now be used to find M at $T_e = T_a = 20$°C. We will assume $r_{Hb} = 300$ s/m for the kangaroo rat and 200 s/m for the pocket mouse (estimate from Figure 7.4 and Table 7.3). Also, assume that $r_e = 50$ s/m. The metabolic rates at 20°C are 88 W/m² for the mouse and 65 W/m² for the rat. The ratio of water produced to water evaporated is $E_{pr}/E_{ev} = 0.1\, M/(5 + 0.02\, M)$. For the rat, the ratio is 1.03, and for the mouse, 1.3. These calculations are crude, but they show that the animals produce enough metabolic water to supply their water requirements without any additional water input. They also show that the more favorable water balance of the mouse is probably the result of the higher metabolic rate it requires to maintain constant body temperature. This metabolic requirement increases as temperature decreases, and is apparently too high during the winter months for the mice to remain active because they hibernate during the winter.

References

7.1 Bernstein, M. H. (1971) Cutaneous water loss in small birds. *Condor 73*:468–469.

7.2 Cena, K. and J. L. Monteith (1975) Transfer processes in animal coats, II. Conduction and convection. *Proc. R. Soc. Lond. B. 188*:395–411.

7.3 Kerslake, D. McK. (1972) *The Stress of Hot Environments.* London: Cambridge University Press.

7.4 Kreith, F. (1965) *Principles of Heat Transfer.* Scranton, Pa. International Textbook Co.

7.5 Lasiewski, R. C., M. H. Bernstein, and R. D. Ohmart (1971) Cutaneous water loss in the Roadrunner and Poor-will. *Condor 73*:470–472.

7.6 MacMillen, R. E. and E. A. Christopher (1975) The water relations of two populations of noncaptive desert rodents. *Environmental Physiology of Desert Organisms.* (N. F. Hadley, ed.) New York: John Wiley.

7.7 Monteith, J. L. (1973) *Principles of Environmental Physics.* New York: American Elsevier.

7.8 Morhardt, S. S. and D. M. Gates (1974) Energy-exchange analysis of the Belding ground squirrel and its habitat. *Ecol. Monogr. 44*:17–44.

7.9 Porter, W. P. and D. M. Gates (1969) Thermodynamic equilibrium of animals with environment. *Ecol. Monogr. 39*:245–270.

7.10 Robinson, D. E., G. S. Campbell, and J. R. King (1976) An evaluation of heat exchange in small birds. *J. Comp. Physiol. 105*:153–166.

7.11 Schmidt-Nielsen, K. (1969) The neglected interface: the biology of water as a liquid-gas system. *Quart. J. Biophys. 2*:283–304.

7.12 Schmidt-Nielsen, K. (1972) *How Animals Work.* London: Cambridge University Press.

7.13 Schmidt-Nielsen, K., J. Kanwisher, R. C. Lasiewski, J. E. Cohn, and W. L. Bretz (1969) Temperature regulation and respiration in the Ostrich. *Condor 71*:341–352.

7.14 Scholander, P. F., V. Walters, R. Hock, and L. Irving (1950) Body insulation of some arctic and tropical mammals and birds. *Biol. Bull. 99*:225–236.

Problems

7.1 Is a migrating bird's range limited by water storage or food storage? Do your calculations for a 0.027 kg sparrow. Assume M for flight $= 6M_b$. Stored water (including metabolic water) is about 15 percent of body mass or 4 g. Stored fat is also $\sim$ 4g. Energy content of fat is 40 kJ/g. Also assume $T_a = 10°C$, $\rho_{va} =$ 4 g/m^3, $T_{skin} = 38°C$.

7.2 Make a climate diagram for an animal.

7.3 How much food does a 500-kg caribou need (kg/day) to survive an arctic winter with average T_a of $-20°C$? Assume $u = 3$ m/s.

7.4 What is the equivalent temperature for a sunbather on a beach at noon on a clear day, with $T_a = 30°C$, $u = 2$ m/s, $a_s = 0.7$?

8 Humans and Their Environment

Human-environment interaction involves the same principles we discussed in Chapter 7. However, we need to look at three additional factors. These are (a) the role of clothing, (b) latent heat loss from sweating, and (c) environments that are comfortable. These can be examined as we consider survival in cold environments, survival in hot environments, and comfort. The variables that we need to consider are metabolic rate, surface area, latent heat exchange, body temperature, and body (clothing and tissue) resistance.

Area, Metabolic Rate, and Evaporation

The total body area in square meters (often called the DuBois area in honor of DuBois and DuBois [1] who first proposed the formula) can be calculated from

$$A = 0.2\ m^{0.425} h^{0.725}$$

where m is the body mass in kilograms and h is the height in meters. As a rough rule of thumb, one can estimate body area for adults from

$$A \simeq 0.026\ m$$

Metabolic rates can be calculated using Equation 7.4, but a better guide can be obtained from measurements. Table 8.1 gives values of M for various activity levels. These activity levels conform quite well to our rules of thumb of M_b = 30–50 W/m² and $M_{max} = 10\ M_b$. To relate these numbers to food intake, they can be multiplied by 1.7 to get kcal/hr for a person with a 2 m² surface area. Thus, desk work would con-

Table 8.1 Metabolic heat production for humans

Activity	M (W/m²)
Sleeping	50
Awake, resting	60
Standing	90
Working at a desk or driving	95
Standing—light work	120
Level walking at 4 km/hr or moderate work	180
Level walking at 5.5 km/hr or moderately hard work	250
Level walking at 5.5 km/hr with a 20-kg pack or sustained hard work	350
Short spurts of very heavy activity such as in climbing or sports	600

Data from Landsberg [8.4].

sume 160 kcal/hr and sleeping 85 kcal/hr. For 8 hr of sleep and 16 hr awake, the daily caloric requirement would be around 3200 kcal, which is generally considered to be a normal requirement for an 80-kg person. If one performed hard physical labor for 12 hr/day and rested for the remaining 12 hr, his caloric intake would need to increase to 6000 kcal/day. For those who exercise for weight control, an hour's strenuous exercise is worth about 600 kcal in excess food intake. If we assume the caloric content of fat to be 40 kJ/g, strenuous exercise for 1 hr would use 63 g of fat. One might conclude from this that regulation of caloric intake is an easier mode of weight control than exercise.

As a note of caution, we need to remember that the values in Table 8.1 are for thermoneutral temperatures. If additional metabolic energy is required for thermoregulation (Equation 7.15) this must be added to the values in Table 8.1.

Latent heat is lost through respiration and water loss directly from the skin. In Chapter 7 we derived an expression for respiratory latent heat loss, and found it to be around 0.08 M in relatively dry environments (Equation 7.5). In more moist environments, it would be smaller. Evaporation from the skin in the absence of thermal sweating is called insensible perspiration, and can be calculated from Equation 7.6 using the appropriate value for skin resistance, r_{vs}, from Table 7.2. For $\rho_{va} = 7$ g/m³ and $r_{vs} = 7.7$ ks/m, $\lambda E_s = 9$ W/m². This is about twice the respiratory latent heat loss at $M = M_b$.

The core temperature of the body depends mainly on metabolic heat production until environmental conditions become too severe for proper thermoregulation. A convenient equation expressing the relationship between metabolic rate and core temperature is [8.3]

$$T_b = 36.5 + 4.3 \times 10^{-3} M$$

where M is in W/m^2.

Resistance to heat transfer in the human body is, as with other homeotherms, subject to vasomotor control. The tissue resistance (r_{Ht}) is varied, within limits, to balance the energy budget. The limits given in Table 7.3 are $r_{Ht} = 90$ s/m for vasoconstriction and $r_{Ht} = 15$ s/m for vasodilation. These values were calculated from Kerslake ([8.3], Figure 7.22). Monteith gives a range of 120 to 30 s/m. The difference is probably due to acclimatization of subjects or possibly subject-to-subject variation. In any case, we will use the range 90 to 15 s/m for our calculations.

Clothing resistance for humans is more difficult to treat than coat resistance for animals because it can be extremely variable and because it is usually strongly windspeed dependent. Normal indoor clothing has a resistance of around 100 s/m in still air. In moving air, this is drastically reduced, as common experience will verify. In the absence of resistance measurements for a given assemblage of clothing, one can use estimates based on windspeed, permeability, thickness, and ventilation of the clothing.

Survival in Cold Environments

Equation 7.15 will be used as the basis for our examination of energy and thermal resistance requirements for humans. Consider first the lowest temperature at which a human can survive. This can be found by assuming some values for r_{Ha}, λE, and r_{Hb}. If we assume $d = 0.17$ m, $u = 3$ m/s, $\lambda E_R = 0.1\ M$, $\lambda E_s = 9$ W/m^2, and $T_b = 36$°C then the lowest equivalent temperature for survival can be calculated for various resistances and metabolic rates. The resistances needed for Equation 7.15 are the radiative resistance, r_r, the convective resistance, r_{Ha}, and the parallel combination of the radiative and convective resistances, r_e. From Table A.3, with $T = 0$°C, $r_r = 260$ s/m.

$$r_{Ha} = 0.7 \times 307\left(\frac{d}{u}\right)^{1/2} = 215\left(\frac{0.17}{3}\right)^{1/2} = 51 \text{ s/m}$$

$$r_e = \frac{r_r r_{Ha}}{r_r + r_{Ha}} = \frac{260 \times 51}{260 + 51} = 43 \text{ s/m.}$$

These values and those previously assumed are substituted into Equation 7.15 and it is solved for equivalent blackbody temperature to give

$$T_e = 36 - \frac{(0.9\ M - 9)\ (r_{Hb} + 43)}{1200} \tag{8.1}$$

Values for T_e as a function of M are plotted in Figure 8.1 using $r_{Hb} = 90$, 300, and 1000 s/m.

The lowest value is for no clothing, the second is for a wool business suit, and the third would be for thick winter clothing covering most of the body. We see that survival is possible at quite low temperatures, even without clothing, if metabolic rate can be kept high. Darwin [8.1] makes some interesting observations of survival among the natives of Tierra del Fuego under these conditions:

> The climate is certainly wretched: the summer solstice was now passed, yet every day snow fell on the hills, and in the valleys there was rain accompanied by sleet. The thermometer generally stood about 45°F., but in the night fell to 38° or 40° . . . While going one day on shore near Wallaston Island, we pulled alongside a canoe with six Fuegians. These were the most abject and miserable creatures I anywhere beheld. On the East coast the natives, as we have seen, have guanaco cloaks, and on the west, they possess sealskins. Amongst these central tribes the men generally have an otter skin, or some small scrap about as large as a pocket handkerchief, which is barely sufficient to cover their backs as low down as their loins. It is laced across the breast by strings, and according as the wind blows, it is shifted from side to side. But the Fuegians in the canoe were quite naked, and even one full-grown woman was absolutely so. It was raining heavily, and the fresh water, together with the spray, trickled down her body. In another harbor not far distant, a woman, who was suckling a recently-born child, came one day alongside the vessel, and remained there out of mere curiosity, whilst

FIGURE 8.1. Minimum equivalent temperature for survival at various metabolic rates and clothing plus tissue resistances.

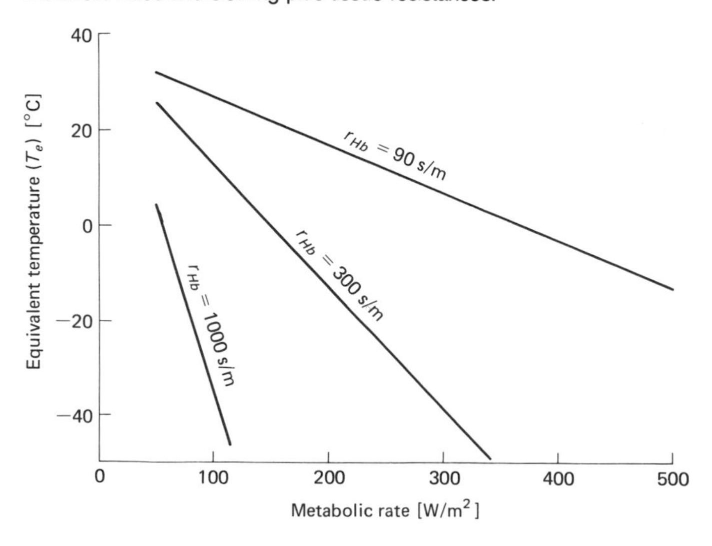

the sleet fell and thawed on her naked bosom, and on the skin of her naked baby! . . . At night, five or six human beings, naked and scarcely protected from the wind and rain of this tempestuous climate, sleep on the wet ground coiled up like animals.

If we were to plot Equation 8.1 in a different way to show r_{Hb} as a function of T_e, we could find the clothing thermal resistance required for any given activity and environment. This is shown in Figure 8.2. This figure is intended to provide information similar to a chill factor table, but is much more useful since it takes into account activity (through M) and radiation (through T_e). To illustrate the use of Figure 8.2, assume you were to go out when $T_e = -20°C$. If you intended to stand for long periods of time, you would need $r_{Hb} = 830$ s/m. For running, you would only need $r_{Hb} = 150$ s/m. The effect of wind on these numbers is relatively small so the graph shows only one windspeed (2 m/s). However the effect of wind on clothing resistance can be large and must be taken into account when choosing the amount of clothing necessary to provide the required r_{Hb}. It is the effect of wind on r_{Hb} that is primarily responsible for the chill factor. A rough idea of wind effects on clothing was obtained by reanalyzing some available data on vapor transport through layers of clothing material [8.5]. The ratio of clothing thermal resistance in wind to resistance with no wind was found to be

FIGURE 8.2. Thermal resistance required for survival in cold at various equivalent temperatures and activity levels.

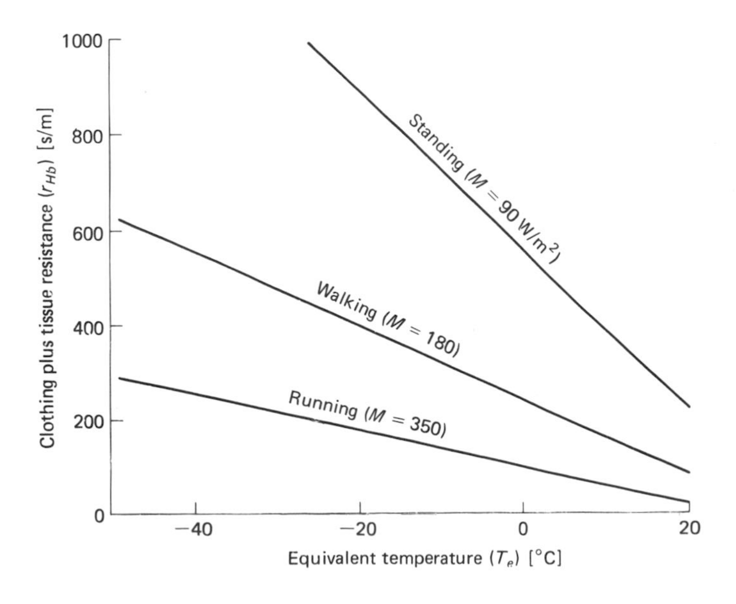

$$r_{Hc}/r^{\circ}_{Hc} = 1 - 0.05\ P^{0.4}\sqrt{u} \qquad (8.2)$$

where P is the air permeability of the clothing (Table 8.2). Clothing efficiency in wind is plotted in Figure 8.3 for various air permeabilities.

Survival in Hot Environments

The same considerations apply to survival in hot environments as did for the upper lethal limit line in the climate diagram for animals. However, we need to look at one additional factor—that of sweating. The rate of sweat evaporation may be either environmentally or physiologically controlled. If the skin surface is wet, rate of water loss from sweating is given by Equation 7.6 with $r_{vs} = 0$. If the skin surface is not wet, latent heat loss is controlled by sweat rate. Control of sweat rate is still not entirely understood, but apparently it involves sensing of surface heat flux [8.3]. Thus, changes in metabolic rate or external environment can cause changes in sweat rate. This seems reasonable, since the requirement for body temperature control is that the heat budget be balanced. Maintaining a balanced heat budget as metabolic rate or equivalent temperature increase, without increasing body temperature, requires increased heat dissipation. There appears to be a hierarchy of physiologic responses to increase dissipation. As heat load increases, tissue thermal resistance decreases first, then sweating begins. Finally, body temperature begins to rise. All of these responses are accompanied by increases in skin temperature. Attempts have been made to relate the first two responses to skin temperature, with only partial success. Body temperature begins to increase when skin temperature reaches about

Table 8.2 Air permeabilities of representative fabrics

Fabric	Air permeability
Very open weave shirt	395
Knit cotton undershirt or T shirt	220
13 civilian shirts (broadcloth or Oxford weave):	
Range	233–24
Average	93
Light worsteds, gabardines, tropicals: Range of 6	60–42
Seersucker suiting: Range of 5	41–50
Uniform twill, 8.2 oz., Army	12
Poplin, 6 oz., Army	6
Byrd cloth, wind resistant	3
J0 cloth, special wind resistant	0.9

Data from Newburgh [8.5].

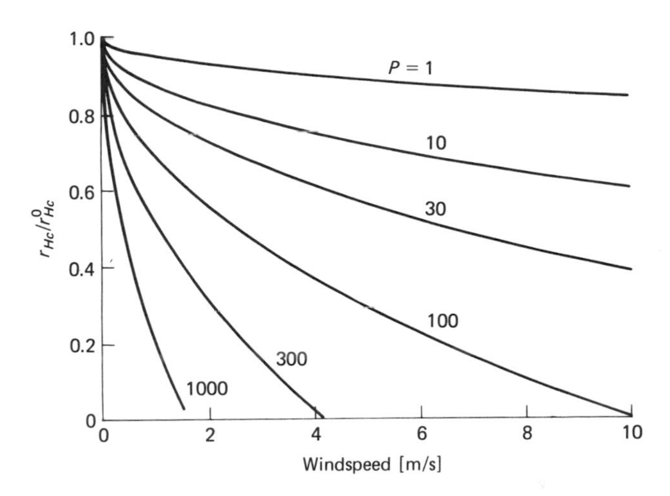

FIGURE 8.3. Clothing effectiveness in wind as a function of wind speed and clothing air permeability. (Equation 8.2.)

36°C. One becomes extremely uncomfortable at skin temperatures of 36°C or greater.

Since we are primarily interested in the highest heat loads that can be sustained by humans at steady state, we need only consider the maximum rate of sweating that can be sustained. Maximum sweat rate is extremely variable, depending on acclimatization of subjects and duration of exposure. Rates as high as 4 kg/hr have been reported for short time periods [8.5]. Typical rates are much lower than this, and a value commonly used for heat stress calculations is around 1 kg/hr. The latent heat flux equivalent to this is around 380 W/m^2, which we will take as the maximum rate of sweating for an average person. For a heat-stressed person, the latent heat loss will therefore be either 380 W/m^2 (physiologically limited, skin remains dry) or the value predicted by Equation 7.6 with $r_{vs} = 0$ (environment limited, skin remains wet), whichever is smaller. This is conveniently shown in graphical form. Figure 8.4 gives λE_s as a function of clothing plus boundary layer resistance and atmospheric vapor density. Boundary layer resistance for vapor transport is given by Equation 6.19. For porous clothing in which both heat and water vapor are transported mainly by free and forced convection we can assume as a very rough approximation $r_{vc} \simeq r_{Hc}$. In a 3-m/s wind, boundary layer resistance is $r_{va} = 0.7 \times 283\ (0.17/3)^{1/2} = 47$ s/m. If we assume $r_{vc} = 100$ s/m, then with $\rho_{va} = 5$ g/m^3 Figure 8.4 shows that the water supply rate would limit latent heat loss to 380 W/m^2, but if $\rho_{va} = 20$ g/m^2 the environment would limit it to 330 W/m^2.

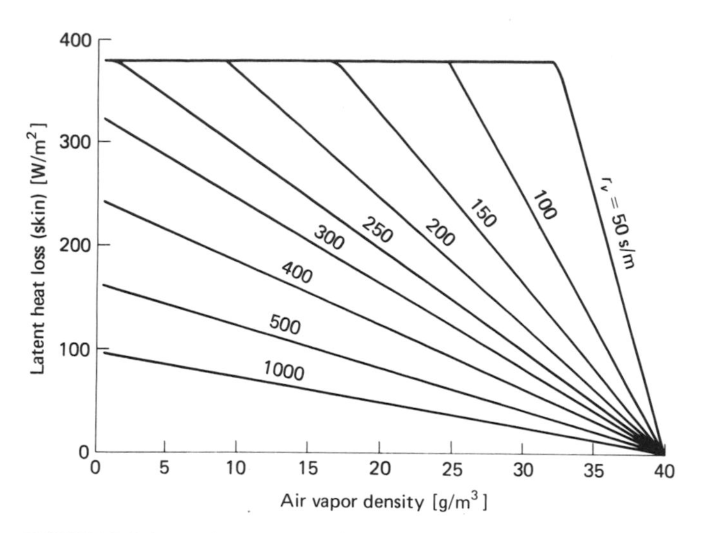

FIGURE 8.4. Latent heat loss from the skin of heat-stressed humans as a function of atmospheric vapor density and boundary layer plus clothing vapor diffusion resistance.

Survival under heat-stress conditions can be predicted using the energy budget equation, but we need to rederive it without the assumption that λE_s is combined with λE_R, as it is in Equation 7.15. This is most easily done by drawing an equivalent electrical circuit with thermal resistances being represented by electrical resistances, heat concentrations by voltages, and heat flux densities by current sources or sinks. The result of such an analysis [8.6] is

$$\frac{\rho c_p(T_b - T_e)}{r_{Hb} + r_e} = M - \lambda E_R - \lambda E_s \frac{(r_{Hc} + r_e)}{(r_{Hb} + r_e)}. \quad (8.3)$$

When $\lambda E_s = 0$ or $r_{Hc} >> r_{Ht}$, this equation reduces to Equation 7.15.

As an example of the use of Equation 8.3, we can investigate the effect of clothing on the maximum equivalent temperature that can be tolerated by a person working at various rates. We will assume $\rho_{va} = 5$ g/m³, $T_b = 38°$C, $r_e = 40$ s/m, $r_{Ht} = 15$ s/m, $r_{Hc} = r_{vc}$, $\lambda E_R = 0.1\ M$, and $r_{va} = 47$ s/m. Results of the calculations are shown in Figure 8.5. Since environmental equivalent temperature is higher than body temperature, adding coat resistance reduces the heat load. Latent heat loss is not reduced under these conditions because it is physiologically limited by water supply rate. The inflection point of the graph occurs when coat resistance becomes large enough to start controlling water loss. Thus for these conditions, light clothing would reduce heat load. We need to keep in mind, though, that

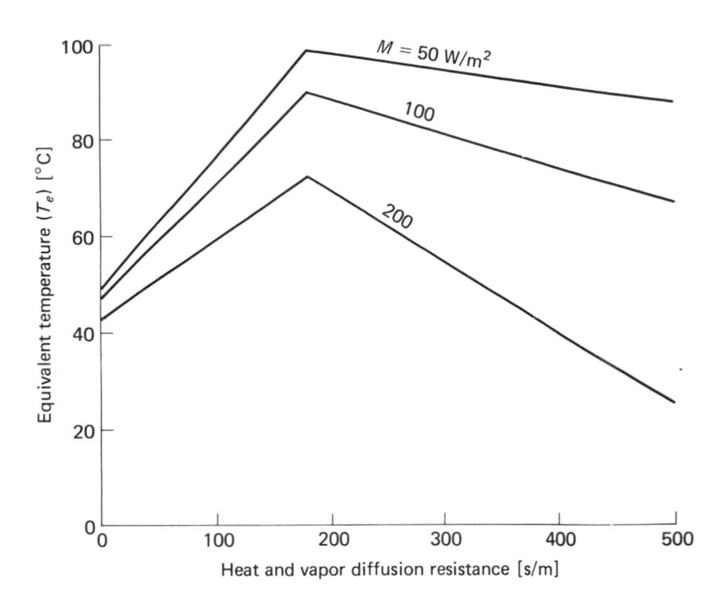

FIGURE 8.5. Maximum tolerable equivalent temperature as a function of clothing resistance for an air vapor density of 5 g/m^3 and wind of 3 m/s at three metabolic rates.

this would not be true at higher vapor densities where any increase in coat resistance would reduce latent heat loss. If the atmospheric vapor density is high enough to keep the skin wet without clothing, then any addition of clothing will decrease dissipation of heat. This brings out the point that clothing must be matched to environment to be most useful in minimizing heat stress. Proper clothing for one hot environment would not necessarily be proper clothing for another.

Equivalent Wet Blackbody Temperature

Much effort has gone into deriving a single measurement that will indicate environmental heat stress for humans [8.3]. The environmental variables that affect heat stress are radiation, temperature, vapor density, and diffusion resistances to heat and vapor. For cold stress, where latent heat loss could be treated as a fixed value relatively independent of environment, the equivalent blackbody temperature adequately combined the radiation and heat transfer characteristics of the environment into a single number. An appropriate energy budget equation was then used to indicate the "strain" imposed by a given "stress," the stress being indicated by the equivalent temperature. It would seem reasonable to attempt to extend the equivalent temperature concept to include atmospheric vapor density. If this could be done, we would again be able to combine all relevant environmental variables into a "stress index," and with an appropriately derived energy budget equation, could indicate the resulting strain on the individual.

The derivation proceeds in a way similar to the derivation of the equivalent blackbody temperature. Substituting Equation 7.6 for λE_s and Equation 7.17 for T_e into Equation 8.3 gives an energy budget equation in terms of physiologic and environmental variables. The vapor density difference in Equation 7.6 can be approximated using the Penman transformation (to be discussed in detail in Chapter 9) to give

$$\begin{aligned} \rho'_{vs} - \rho_{va} &= (\rho'_{vs} - \rho'_{va}) + (\rho'_{va} - \rho_{va}) \\ &\simeq s(T_b - T_a) + (\rho'_{va} - \rho_{va}) \end{aligned} \tag{8.4}$$

where s is the slope of the saturation vapor density curve at skin temperature, and skin temperature is assumed to be close enough to body temperature to introduce negligible error by substituting T_b for skin temperature. With these substitutions, the energy budget equation becomes

$$\left(1 + \frac{s}{\gamma^*}\right)(T_b - T_a) - r_e \frac{(R_{\text{abs}} - \epsilon\sigma T_a^4)}{\rho c_p} = \frac{(M - \lambda E_R)(r_{Hb} + r_e)}{\rho c_p} - \frac{(\rho'_{va} - \rho_{va})}{\gamma^*} \tag{8.5}$$

where $\gamma^* = \gamma r_v/(r_{Hc} + r_e)$ and $r_v = r_{vs} + r_{vc} + r_{va}$. For a person in a saturated blackbody cavity, with $T_{ew} = T_a$ and $\rho_{va} = \rho'_{va}$, the energy budget equation reduces to

$$\frac{\gamma^*}{s+\gamma^*}(M - \lambda E_R) = \frac{\rho c_p (T_b - T_{ew})}{r_{Hb} + r_e}. \tag{8.6}$$

The required definition of equivalent wet blackbody temperature is obtained by subtracting Equation 8.6 from 8.5 to give

$$T_{ew} = T_a + \frac{\gamma^*}{s + \gamma^*}\left[\frac{r_e(R_{\text{abs}} - \epsilon\sigma T_a^4)}{\rho c_p} - \frac{(\rho'_{va} - \rho_{va})}{\gamma^*}\right]. \tag{8.7}$$

It is apparent that these equations are more general forms of Equations 7.15 and 7.17 since, as r_{vs} becomes large, γ^* becomes large, and the equations simplify to the dry equivalent temperature equations. It is also apparent that Equation 8.7 is the heat stress index we were seeking, since it combines all of the relevant environmental variables into a single equivalent temperature. As with the dry equivalent temperature, T_{ew} is the temperature of a poikilotherm ($M - \lambda E_R = 0$) exposed to the same environment as a person, and with identical radiative, convective, and latent heat exchange properties. The temperature of a copper sphere covered with black, moistened cloth, and filled with water, has been used to determine "wet globe temperature," and these measurements have been related to human comfort in hot environments. The exchange properties of such a system would not be identical to those for a human, so the temperature measured in this way would not be the wet equivalent temperature. It is possible, though, that wet globe temperature would correlate with T_{ew}.

When there is some air movement, the skin surface is wet, and clothing vapor and heat transfer resistances can be assumed equal, γ^* can be taken as equal to γ. At body temperature, the term $\gamma/(s+\gamma)$ (Appendix Table A.3) becomes 0.17. The second term in Equation 8.7 is therefore quite small and T_{ew} is usually only a few degrees different from T_a.

If vapor and heat resistances are equal and the skin surface is wet, we can find the maximum equivalent temperature that can be tolerated for various body heat transfer resistances and metabolic rates. For these calculations we will take body temperature as 38°C, r_e as 40 s/m and λE_R as 0.05 M. Figure 8.6 shows the results of these calculations. As expected, maximum T_{ew} decreases as M increases and as r_{Hb} increases. The wet equivalent temperature must always be lower than body temperature to maintain a steady-state energy balance.

FIGURE 8.6. Maximum tolerable wet equivalent temperature at various tissue plus coat resistances as a function of metabolic rate.

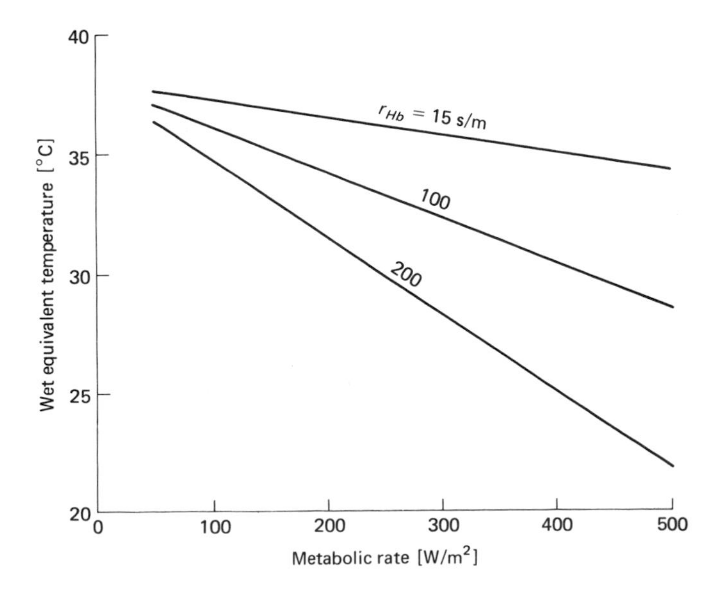

Comfort In their usual activities, humans are generally not so concerned with minimum conditions for survival as they are with comfort. For this, the energy budget approach still gives the needed answers. We need only put in physiologic parameters that we think represent comfort. These, of course, vary considerably from individual to individual. For our purposes we will assume that a person is comfortable if $T_b = 37°C$, and $r_{Ht} = 40$ s/m. For normal indoor conditions we will take r_{Ha} as 200 s/m and r_r as 200 s/m, so $r_e = 100$ s/m. Thermal resistance of normal indoor clothing will be taken as 100 s/m. Combining Equations 7.5 and 7.6, and assuming both skin and expired air temperature are 34°C, gives

$$\lambda E = (2.3 \times 10^{-3}M + 0.31)(37 - \rho_{va}).$$

When this expression is substituted into Equation 7.15 we obtain estimates of comfortable temperature. These are plotted in Figure 8.7 for $\rho_{va} = 5$ and 25 g/m³.

If the room wall temperature is equal to air temperature, then $T_e = T_a$. Figure 8.7 shows that for normal active metabolic activity ($M = 90$), a comfortable room temperature at low vapor density would be 22°C, as we would expect. In a humid room, the comfortable temperature would be 19°C. Thus it is possible to reduce room temperature and maintain

FIGURE 8.7. Comfortable equivalent temperature for two air vapor densities as a function of metabolic rate.

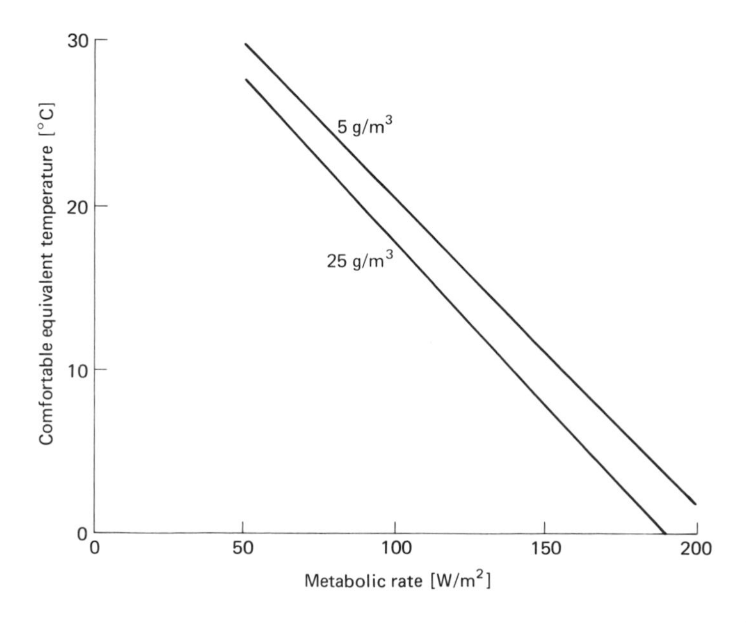

comfort if the air is humidified. This has been suggested as a means for reducing heating costs. A more complicated analysis would be necessary to determine whether humidifying the air would actually reduce heating costs since one would need to compare the cost of evaporating the water to humidify the air with the cost of keeping the room a few degrees warmer.

Figure 8.7 also shows that a relatively small change in activity results in a fairly large change in comfortable temperature. This is confirmed by common experience.

Many other aspects of comfort could be investigated using the energy budget equation. For example, one sometimes feels cold in a room even when the thermometer indicates an air temperature of 22°C. Quite often, wall temperature measurements show temperatures significantly below air temperature. A calculation of T_e from Equation 7.17, assuming T_{wall} = 15°C and r_e = 100 s/m, gives T_e = 18.7, a quite uncomfortably low temperature for a resting person. Thus we see that our radiant energy environment is very important to comfort. Rooms with large window areas may be quite uncomfortable in very cold or very hot weather, even if room air temperature is controlled at desirable levels.

References

8.1 Darwin, Charles (1832) *Journal of Researches into the Natural History and Geology of the Countries Visited During the Voyage of H.M.S. Beagle Round the World.* London: John Murray. (Quoted in Reference [8.5], Ch. 1.)

8.2 Dubois, D. and E. F. Dubois (1915) The measurement of the surface area of Man. *Arch. Intern. Med. 15*:868–881.

8.3 Kerslake, D. McK. (1972) *The Stress of Hot Environments.* London: Cambridge University Press.

8.4 Landsberg, H. E. (1969) *Weather and Health, an Introduction to Biometeorology.* Garden City, N.Y.: Doubleday.

8.5 Newburgh, L. H. (ed.) (1968) *'Physiology of Heat Regulation and The Science of Clothing.* New York: Hafner.

8.6 Robinson, D. E., G. S. Campbell, and J. R. King (1976) An evaluation of heat exchange in small birds. *J. Comp. Physiol. 105*:153–166.

Problems

8.1 Find your body surface area using the DuBois formula and compare the result to the area predicted by A = 0.026 m

8.2 What clothing would you recommend for a walk on a windy, overcast day (u = 10 m/s, T_a = 0)? Show calcualtions on which you base the recommendation.

8.3 What is the hottest wet equivalent temperature you could work at with u = 3 m/s and light clothing?

8.4 What is the wet equivalent temperature for the sunbather in Problem 7.4 if $\rho_{va} = 25$ g/m^3? What rate of water consumption would be required to maintain water balance?

9 Plants and Their Environment

In our previous discussions of animals and humans we have found that certain environments are unsuitable for life. An animal can choose its environment to best suits its energetics. Plants are not able to move around to find a suitable environment, but we see considerable evidence that selection and adaptation result in leaf morphologies, canopy structures, etc. which give the plants native to a given environment a competitive advantage for that location. Desert plants, for example, tend to have quite narrow leaves, while leaves of plants from more moist environments may be much larger. We might ask ourselves what environmental limitations there are to leaf size and other leaf characteristics related to energy exchange or whether there is an optimum leaf form for a particular leaf environment. Obviously, factors outside the scope of this discussion will influence the optimum leaf form for a given environment, such as structural economy, canopy structure, and predation. We will concern ourselves primarily with the effects of physical environment on photosynthesis and transpiration with the idea of indicating what leaf characteristics might be best suited to a particular environment.

Two factors must be favorable for a leaf to remain alive. Average net photosynthesis must be positive, and it must remain at nonlethal temperatures. If net photosynthesis is negative for a period, the leaf abscises because there is no mechanism for importing sugars to mature leaves. Net photosynthesis is determined by environmental factors and by the water balance of the leaf. We will first consider the effects of environment on leaf photosynthesis and water loss. We will then use the energy budget approach to determine the leaf temperature as a function of environmental variables. Finally we will use the leaf temperature and photosynthesis models to indicate optimum leaf form for a specified environment.

Photosynthesis

Numerous models have been proposed that relate photosynthesis to environmental variables. We will use one described by Lommen *et al.* [9.1]. They have two photosynthesis models, one for respiring leaves and the other for nonrespiring leaves. The model for nonrespiring leaves is the simplest and will be used here. Respiration could be added to this model, as Lommen *et al.* did, or by subtracting compensation point irradiances and CO_2 concentrations, as was proposed by van Bavel [9.5]. Photosynthesis involves two rate processes: CO_2 diffusion from the atmosphere to the chloroplast and biochemical fixation of the CO_2. Each of these processes has many complex steps, details of which need not concern us here. Under certain conditions (say high light levels), we might expect the biochemical processes to be using up CO_2 as fast as it can diffuse to the chloroplast. Under other conditions (low light or temperature) diffusion might be rapid enough to supply CO_2 faster than it can be used by the chloroplast. With intermediate conditions, photosynthesis might be partially limited by both processes. The diffusion process in photosynthesis is described by Equation 6.3. For CO_2,

$$P = \frac{\rho_{ca} - \rho_{cc}}{r_{ca} + r_{cs} + r_{cm}} \tag{9.1}$$

where P is the negative CO_2 flux density (g $m^{-2}s^{-1}$),[1] ρ_{ca} and ρ_{cc} are the concentrations of CO_2 in air and at the chloroplast, and r_{ca}, r_{cs}, and r_{cm} are the diffusion resistances to CO_2 in the boundary layer, stomates, and mesophyll cells and cell walls. Problems in defining these resistances will be discussed later. For now, we will assume $r_{ca} = 1.4\, r_{va}$ and $r_{cs} = 1.65\, r_{vs}$, where r_{va} and r_{vs} are the boundary layer and stomatal resistances for water vapor. Recent calculations [9.6] indicate that r_{cm} is small compared to the other resistances. We will assume $r_{cm} = 0$. All of the unknowns in Equation 9.1 can be measured, under appropriate conditions, except the concentration of CO_2 at the chloroplast. It depends on the rate at which photosynthesis proceeds. A suitable empirical description of

[1] Photosynthetic rate (P) is considered positive when CO_2 is photosynthetically fixed (CO_2 flux density is negative). Flux densities for leaves are usually computed per unit projected leaf area, rather than averaging over the entire surface as was done with animals. Thus, all measurements are normalized to a unit one-sided leaf area.

this process is given by a Michaelis-Menten-type equation for the overall chemical reaction:

$$P = \frac{P_M}{1 + \frac{K}{\rho_{cc}}} \tag{9.2}$$

where P_M is the rate of photosynthesis at CO_2 saturation and K is a rate constant which equals the chloroplast CO_2 concentration when $P = P_M/2$. Equations 9.1 and 9.2 can be combined and solved for P to give

$$P = \frac{(\rho_{ca} + K + r_c P_M)}{2\, r_c} - \frac{[(\rho_{ca} + K + r_c P_M)^2 - 4\rho_{ca} r_c P_M]^{1/2}}{2\, r_c} \tag{9.3}$$

where the total resistance for CO_2 diffusion is $r_c (= r_{ca} + r_{cs} + r_{cm})$.

The maximum rate of photosynthesis at CO_2 saturation (P_M) will depend on leaf temperature and the flux density of photosynthetically active radiation at the leaf surface. An equation that gives P_M as a function of these variables is

$$P_M = \frac{P_{MLT} G(T)}{1 + \frac{K_L}{\text{PAR}}} \tag{9.4}$$

where P_{MLT} is the maximum photosynthesis at light and CO_2 saturation and optimum temperature, PAR is the average 400–700 nm irradiance (W/m^2), K_L is the rate constant for light [irradiance at which $P_M = P_{MLT} G(T)/2$], and $G(T)$ is a dimensionless temperature factor. An empirical equation that describes temperature response of photosynthesis in many species with reasonable accuracy [9.4] is

$$G(T) = \frac{2(T_L + A)^2(T_M + A)^2 - (T_L + A)^4}{(T_M + A)^4} \tag{9.5}$$

where T_M is the temperature for maximum photosynthesis, T_L is leaf temperature, and A is a constant, which is adjusted to best fit photosynthesis data for the species being studied.

Combining Equations 9.3, 9.4, and 9.5, we can describe the effects of environment on photosynthesis. Results are shown in Figure 9.1. Values used in the equations are those shown in Table 9.1, if not otherwise shown on the figure. Figure 9.1a shows that photosynthesis is quite sensitive to temperature at low diffusion resistance, but as r_c increases the

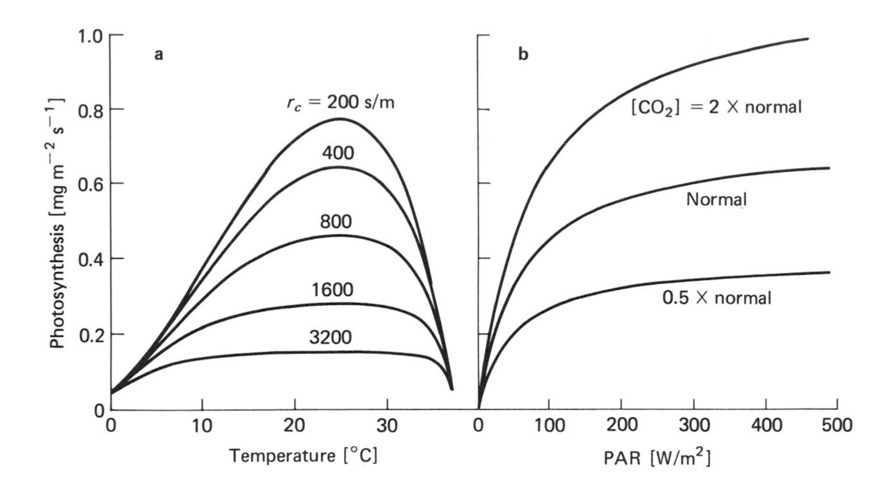

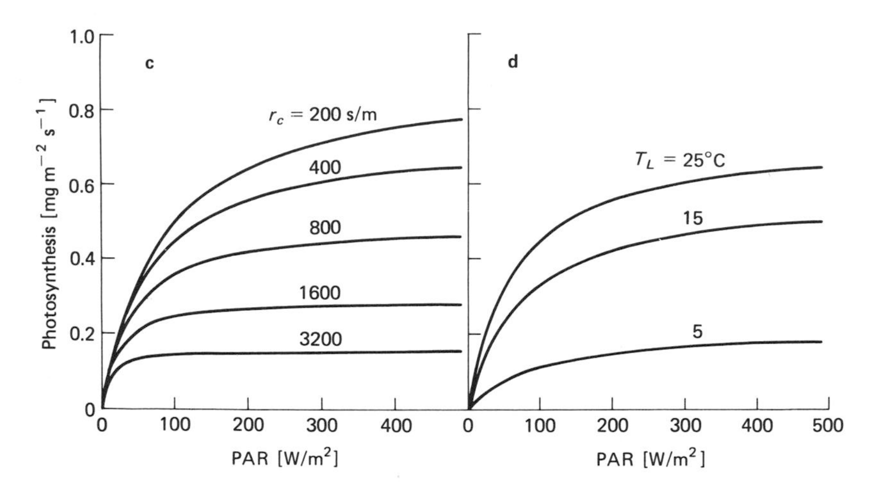

FIGURE 9.1. Photosynthesis as a function of various environmental and physiological variables for a nonrespiring leaf. Values of parameters are given in Table 9.1 unless otherwise specified on the figure.

temperature optimum becomes very broad. In Figure 9.1b, the effect of CO_2 concentration is shown. Light saturation occurs at low light levels with low ρ_{ca}, but is at much higher light levels for high ρ_{ca}. The highest light level shown is PAR =

Table 9.1 Standard parameters for Equations 9.3, 9.4, and 9.5

ρ_{ca} = 0.54 g/m^3 [a]	K_L = 100 W/m^2
K = 0.44 g/m^3	T_M = 25°C
r_c = 400 s/m	A = 5°C
P_{MLT} = 2×10^{-3} g m^{-2}s^{-1}	PAR = 500 W/m^2

[a] Normal ambient atmospheric concentration.

500 W/m^2. This corresponds roughly to full sun direct plus diffuse radiation, averaged over the surface of one side of the leaf. Only about half of the total solar irradiance is photosynthetically useful. Figures 9.1c and 9.1d show temperature and resistance effects on photosynthesis as a function of light. At low temperatures or high resistances, light saturation occurs at relatively low light levels.

From these curves, one gets a feeling for which variables are controlling photosynthesis under a given set of environmental and physiologic conditions. We can use this information to predict which physiologic characteristics are important to photosynthesis and which environments are best suited to photosynthesis.

Transpiration and the Leaf Energy Budget

An understanding of transpiration is important because water loss influences both the water and the energy budget of the plant. The transpiration rate for a leaf is given by Equation 6.2 with E defined as the water vapor flux per one-sided leaf area. The driving force for vapor loss is the vapor density difference between the substomatal cavaties in the leaf (saturation value at leaf temperature) and the ambient air. We need to be a little careful in defining the diffusion resistance for vapor flow since abaxial and adaxial diffusion resistances are seldom equal, and a straightforward averaging process would lead to errors. We need to define a resistance such that the water loss per unit projected leaf area would be the same as the water loss from a real leaf with its measured resistances. Thus,

$$\frac{1}{r_v} = \frac{1}{r_{vs}^{ab} + r_{va}} + \frac{1}{r_{vs}^{ad} + r_{va}} \tag{9.6}$$

where r_{vs} is epidermal (stomate and cuticle) resistance, r_{va} is boundary layer resistance (Equation 6.19) and the superscripts *ad* and *ab* are for adaxial and abaxial surfaces. For a hypostomatous leaf (stomates on one side) $r_v = r_{vs} + r_{va}$. For an amphistomatous leaf with equal abaxial and adaxial resistances, $r_v = (r_{vs} + r_{va})/2$. For our purposes, we will assume the leaf to be amphistomatous with resistances of the two surfaces equal.

The energy budget equation for a leaf can be written as

$$R_n - H - \lambda E = 0 \tag{9.7}$$

where R_n is net radiation, H is sensible heat, and λE is latent heat flux density. We assume that metabolic energy (photosynthesis and respiration) is negligibly small. All three terms in Equaion 9.7 involve leaf temperature. Since we wish to investigate environmental effects on leaf temperature, it would be convenient to find suitable linear approximations for the non-linear functions in R_n and E.

The vapor-density-dependent function can be linearized using a substitution worked out by Penman [9.2]. We will use this again later in deriving the Penman evapotranspiration equation. The substitution is shown in Figure 9.2 and is expressed by the equation

$$\rho'_{vL} - \rho_{va} = \rho'_{va} - \rho_{va} + s(T_L - T_a) \tag{9.8}$$

where s (g m^{-3}K^{-1}) is the slope of the saturation vapor density curve at $(T_L + T_a)/2$; T_L and T_a are leaf and air temperatures; ρ'_{vL} and ρ'_{va} are the saturation vapor densities at leaf and air temperature; and ρ_{va} is ambient vapor density. In practice, s is usually determined at T_a, so some error results, but this is seldom serious. Equation 9.8 gives us the vapor density difference between leaf and air in terms of a temperature difference and a vapor density deficit for the atmosphere. Values for s can be found in Table A.3. If we multiply Equation 9.8 by λ/r_v, we obtain (using Equations 6.1 and 6.2 for H and E)

$$\lambda E = \frac{\lambda(\rho'_{va} - \rho_{va})}{r_v} + \frac{\lambda\, sHr_H}{\rho c_p r_v} \tag{9.9}$$

where $r_H = r_{Ha}/2$ (to define heat boundary layer resistance in terms of projected leaf area). The energy budget equation is $H = R_n - \lambda E$. Substituting and re-arranging we obtain

$$\lambda E = \frac{sR_n + [\rho c_p(\rho'_{va} - \rho_{va})/r_H]}{s + \gamma^*} \tag{9.10}$$

where $\gamma^* = \gamma r_v/r_H$. This equation is known as the Penman equation. It describes transpiration in terms of environmental parameters (R_n, vapor density deficit, temperature) and diffusion resistances. Thus we eliminated leaf temperature from the equation (except for the dependence of R_n on T_L). The real utility of Equation 9.10 will be shown in Chapter 10, but we can use it here for one interesting exercise. We might wonder if increasing the windspeed would increase transpiration. Predicted λE is plotted in Figure 9.3 as a function of wind for two special cases, clear night and full sun. We see that wind increases λE when leaf temperature is below air temperature, but may decrease it when $T_L > T_a$. The reason for the decrease in transpiration in full sun is that the leaf is heated by R_n to a temperature above ambient. The increased wind cools the leaf, decreasing ρ'_{vL}, and therefore decreasing the transpiration. The effect of wind on transpiration is relatively small in full sun with open stomates or in dark with closed stomates. The effect

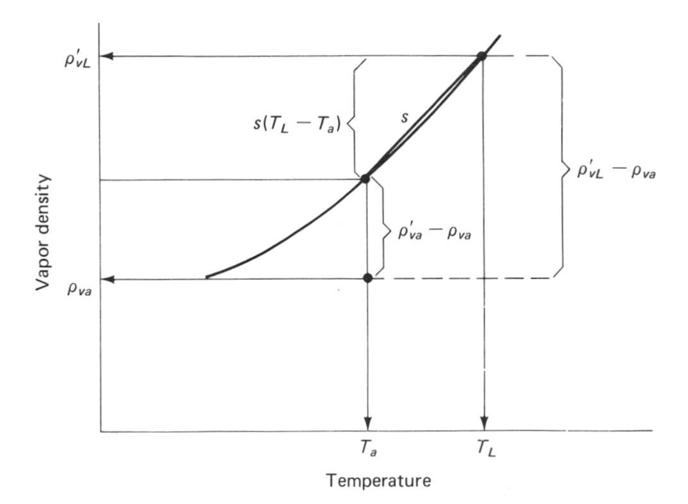

FIGURE 9.2. Vapor density diagram showing the Penman transformation.

is much more pronounced in the dark with open stomates or in full sun with closed stomates.

Leaf Temperature

The temperature of a leaf is determined, as with a poikilothermic animal, by the energy budget of the leaf. We could use the Penman equation and the energy budget equation to determine the leaf temperature as a function of environmental and physiologic variables, but we already derived equations for the temperature of poikilotherms. For a dry system, where latent

FIGURE 9.3. Latent heat loss from a leaf as a function of wind, for selected values of net radiation and diffusion resistance.

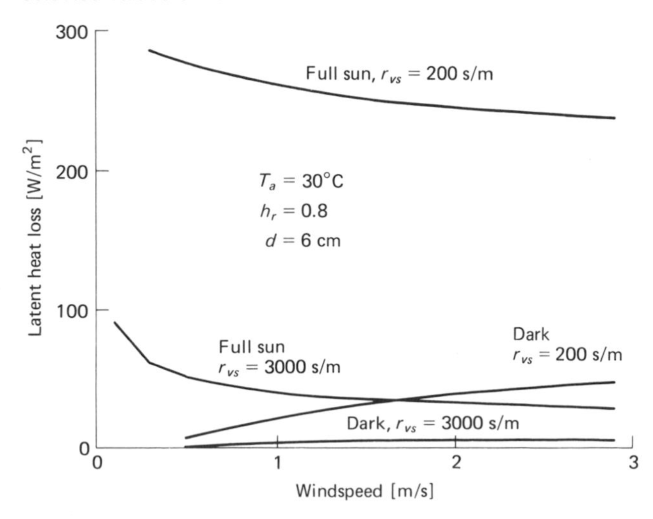

heat exchange was a small and predictable fraction of the total energy budget, the blackbody equivalent temperature (Equation 7.17) gave the temperature of the poikilotherm. For a wet system, where latent heat loss is an important part of the energy budget, the wet equivalent temperature (Equation 8.6) is the temperature of the poikilotherm. For a leaf, observing our convention of using projected area, Equation 8.6 becomes

$$T_L = T_a + \frac{\gamma^*}{s + \gamma^*}\left[\frac{r_e(R_{\text{abs}} - 2\epsilon\sigma T_a^4)}{\rho c_p} - \frac{(\rho'_{va} - \rho_{va})}{\gamma^*}\right] \tag{9.11}$$

where, in this case, $\gamma^* = \gamma r_v/r_e$ and $1/r_e = (1/r_H) + (2/r_r)$. This equation will be useful in determining leaf temperatures from environmental variables so that photosynthetic rates can be calculated.

We can use Equation 9.11 directly to answer a rather significant ecological question. Plant ecologists commonly take air temperature as *the* environmental variable characterizing a site. The implicit assumption is that air temperature and plant temperature are equal, or at least related by some constant, since it is really plant temperature that determines response. We can use Equation 9.11 to check this assumption.

Figure 9.4 shows T_L as a function of leaf characteristic dimension, stomatal diffusion resistance, and absorbed radiation for $T_a = 30°\text{C}$, $h_r = 0.2$ and $u = 1$ m/s. The interesting

FIGURE 9.4. Leaf temperature at various radiation levels and stomatal diffusion resistances.

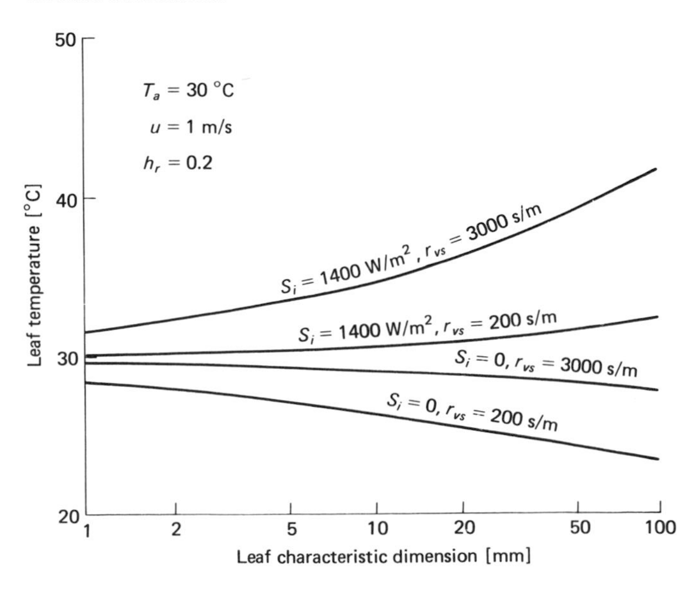

feature of Figure 9.4 is that small leaves remain within a degree or two of air temperature, no matter what the transpiration rate. Large leaves have much higher leaf temperatures when stomates are closed; but when stomates are open, leaf temperatures remain near air temperature. We see from this that small leaves are particularly well suited to hot, dry environments. They can stay cool without evaporation. Some alpine plants, with growth habits such that they are exposed to very little wind, may maintain leaf temperatures well above ambient, but leaves exposed to wind would remain near air temperature.

An effort has been made to sense plant water stress from airplanes or satellites using infrared imaging of the vegetated area. Water stress is inferred from leaf temperature. If we assume that the primary water stress effect is on r_{vs}, we see, from Equation 9.11 and Figure 9.4, that there are many confounding factors including wind, radiation, leaf orientation, and leaf size. Such an approach would probably be unsuccessful on plants with small leaves, on windy days, or at radiation levels much below maximum.

Optimum Leaf Form

With the relationships we have developed, we can now investigate the question of whether there is an optimal leaf form for a plant in a given environment. The approach followed here is similar to the analysis by Taylor [9.3]. We have already seen that small leaves may have an advantage in hot, dry environments because they do not overheat. If we want to maximize photosynthesis for a particular environment, we can calculate leaf temperature from Equation 9.11, then predict photosynthesis using Equation 9.3. Figure 9.5a shows the result for full sun and $T_a = 30°C$. Photosynthesis is plotted as a function of leaf characteristic dimension for several stomatal diffusion resistances. We see that photosynthesis decreases as characteristic dimension increases when r_{vs} is low, and as stomatal diffusion resistance increases, very large leaves are definitely at a disadvantage.

Figure 9.5b is for an ambient temperature of 10°C and full sun. Note that large leaves are an advantage here because they are warmer. Note also that the minimum stomatal diffusion resistance gives little advantage over the one twice as large, again because of cooling of the leaf.

Figures 9.5c and 9.5d show photosynthesis as a function of absorbed short-wave radiation for a leaf with $d = 3$ cm at 30°C and 10°C. At 10°C the preference tends to be for full sun, especially at low r_{vs}. At 30°C the preference is for something less than full sun and little loss is sustained even at very low

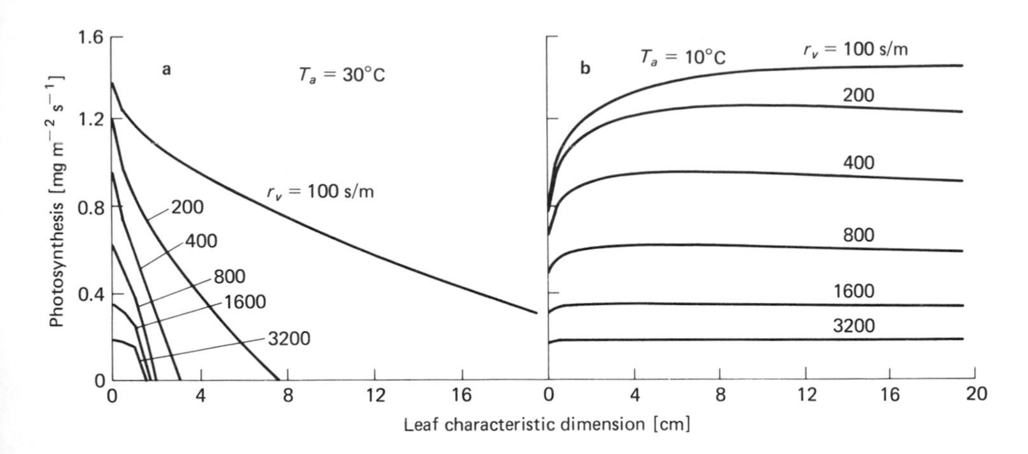

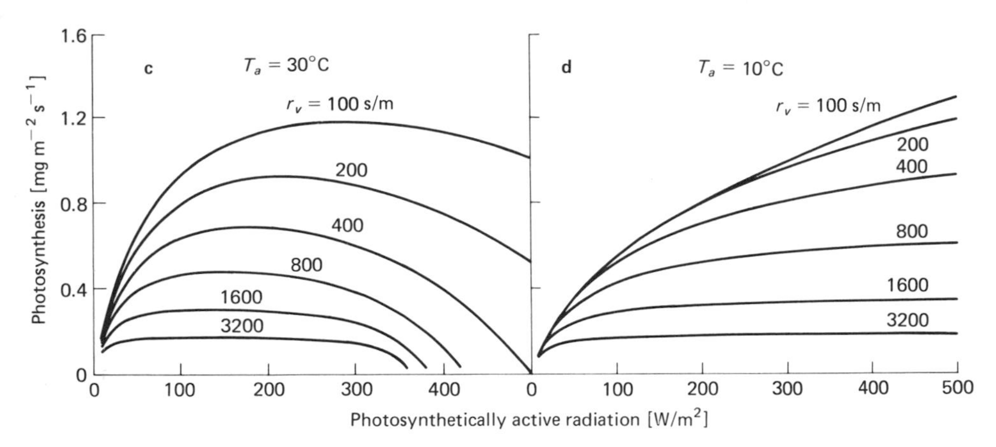

FIGURE 9.5. Photosynthesis as a function of leaf characteristic dimension, irradiance (400–700 nm), and stomatal diffusion resistance (vapor). Photosynthesis parameters from Table 9.1 were used with the following environmental parameters: u = 1 m/s, h_r = 0.5, clear sky. In **a** and **b,** PAR is 500 W/m^2; in **c** and **d,** d = 3 cm.

light levels for closed stomates. There is an indication here that leaf orientation may be more important to the plant than leaf size. Apparently an inclined or erect leaf would have considerable advantage since it would catch full sun in the morning when T_a is low, and would maintain maximum photosynthesis during the heat of the day when the sun is high. This would seem to be a fruitful area for both plant ecology and crop physiology research. It is well known that leaves of many plant species orient themselves with respect to the sun, and even the change in leaf orientation caused by wilting may be an adaptation by some plants to reduce heat load.

There may be conditons under which water use efficiency

(WUE) rather than photosynthesis is to be optimized. We can use Equation 6.2, along with those already mentioned, to investigate WUE as a function of d. Figure 9.6b gives the low-temperature (10°C) response. Small leaves give the highest WUE for all stomatal diffusion resistances. At $T_a = 30°C$, the same is true, but the WUE is more sensitive to d at high r_{vs}.

Figures 9.6c and 9.6d give WUE as a function of absorbed short-wave radiation for a leaf with $d = 3$ cm and $T_a = 10°C$ or 30°C. There is an advantage to intermediate to low radiation levels, especially with high r_{vs}. At 30°C there is a very definite advantage to a radiation level around 10 to 20 percent

FIGURE 9.6. Water use efficiency (mass of CO_2 fixed/mass of H_2O lost) as a function of characteristic dimension, vapor diffusion resistance, and irradiance (400–700 nm). Parameters are the same as those listed for Figure 9.5.

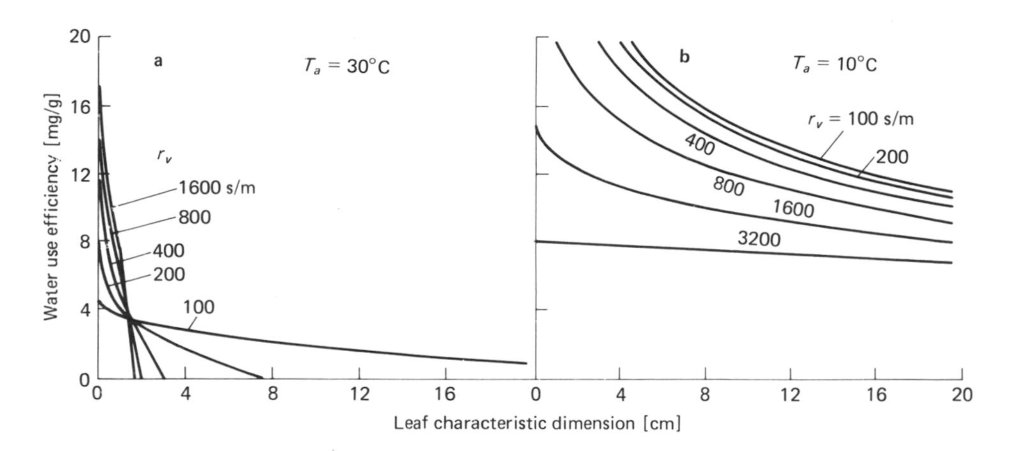

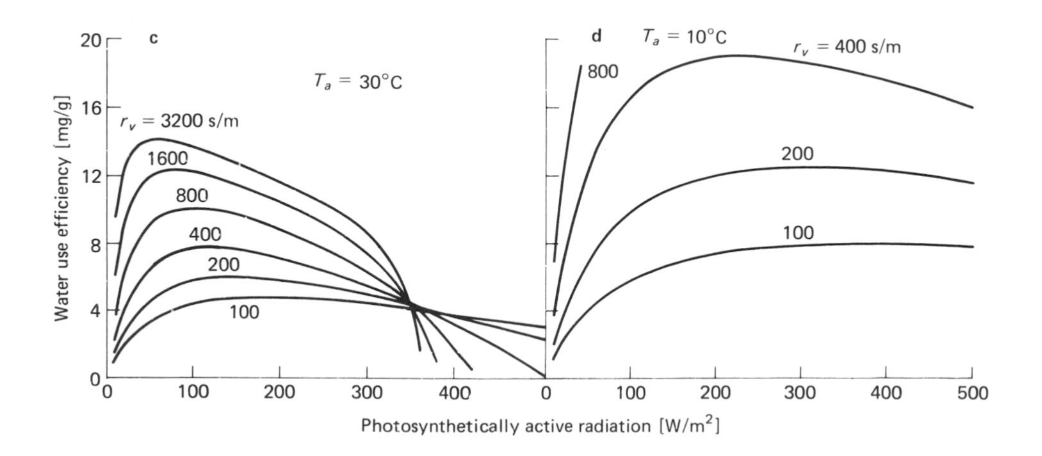

of full sun. Again, it would seem that an erect leaf would have a definite advantage.

One would like to be able to go on and say what characteristics of a plant make it suitable for a given environment, but we don't know that yet. It may be none of those discussed here. Nevertheless, this approach seems promising, and additional information should become available as ecologists apply these ideas. This approach would also appear to have application in crop physiology and crop breeding research.

References

9.1 Lommen, P. W., C. R. Schwintzer, C. S. Yocum, and D. M. Gates (1971) A model describing photosynthesis in terms of gas diffusion and enzyme kinetics. *Planta (Berl.) 98*:195–220.

9.2 Penman, H. L (1948) Natural evaporation from open water, bare soil, and grass. *Proc. R. Soc. A194*:220.

9.3 Taylor, S. E. (1975) Optimal leaf form. *Perspectives in Biophysical Ecology* (D. M. Gates and R. B. Schmerl, eds.). New York: Springer-Verlag.

9.4 Taylor, S. E. and O. J. Sexton (1972) Some implications of leaf tearing in Musaceae. *Ecology 53*:143–149.

9.5 van Bavel, C. H. M. (1975) A behavioral equation for leaf carbon dioxide assimilation and a test of its validity. *Photosynthetica 9*:165–176.

9.6 Yocum, C. S. and P. W. Lommen (1975) Mesophyll resistances. *Perspectives in Biophysical Ecology* (D. M. Gates and R. B. Schmerl, eds.). New York: Springer-Verlag.

Problems

9.1 What is the photosynthetic rate of a 4-cm-wide leaf with PAR = 250 W/m², T_L = 23°C, u = 2 m/s, and r_{cs} = 250 s/m. Use Table 9.1 for other values.

9.2 Use the Penman equation to find the reduction in transpiration rate resulting from treating a leaf with an antitranspirant film. Assume u = 2 m/s, d = 5 cm, R_n = 500 W/m², T_a = 30°C, and h = 0.5. Before treatment r_{vs} = 200 s/m; after treatment r_{vs} = 500 s/m.

9.3 Using Equation 9.11, find the largest difference between leaf and air temperature you would ever expect in nature for a 2-cm-diameter leaf with T_a = 20°C, h_r = 0.5.

9.4 At T_a = 30°C, why do the smallest leaves have the highest photosynthetic rates and the highest water use efficiencies?

10 Exchange Processes in Plant Canopies

In Chapter 9 we examined exchange processes for individual plant leaves and were able to predict photosynthesis and transpiration rate as a function of environmental and plant variables. These results are of limited significance, however, unless we can extend our predictions to plant canopies. Production and water use are determined by the response of the entire canopy rather than that of isolated individual leaves.

We will first look at short-wave radiation within the plant canopy, and then discuss various approaches that have been used to describe loss of water vapor from crop canopies. The importance of the first exercise is obvious since both photosynthesis and transpiration are closely tied to radiant energy. Even some animal studies may require knowledge of radiant energy distribution within or below the canopy.

Radiation in Plant Canopies

Radiant energy within a plant canopy comes from the sun, the sky, the ground, and from canopy elements. A complete analysis of all components would be extremely difficult. Here we will limit our discussion to penetration of canopies by direct and diffuse short-wave radiation, and indicate approaches for considering some other components. Our analysis will be similar to that of Monteith [10.8]. A different approach, which is more useful under some circumstances, is presented by Fuchs [10.2].

Our approach will be to first consider a hypothetical canopy in which opaque leaves are randomly distributed, but with horizontal inclination. The total leaf area index of the canopy (projected leaf area per unit ground surface area) is L_t. If we divide the canopy into many small layers, such that there are no overlapping leaves in any layer, then the average direct short-wave radiation on a horizontal surface below any layer of leaves (S_b) will depend on the value of S_b above the layer and the fractional leaf area in the layer. Thus we can write

$$dS_b = -S_b dL \tag{10.1}$$

where dS_b is the change in S_b that occurs over an increment in leaf area index, dL, and the minus sign indicates that S_b is decreasing as leaf area index (L) increases, since we measure L from the top of the canopy. Equation 10.1 can be integrated to find S_b at any level in the crop:

$$S_b = S_b^0 \, e^{-L} \tag{10.2}$$

where S_b^0 is the value of S_b above the crop. This is similar in form to Bouguer's law, but is really telling us something quite different. Although S_b is the average irradiance below leaf area index, L, we need to remember that this average is made up of spots with irradiance S_b^0, and spots with zero irradiance (since we are assuming diffuse and transmitted short wave are zero). The ratio

$$\frac{S_b}{S_b^0} = e^{-L} \tag{10.3}$$

is therefore the fraction of the total area below L that is sunlit. We can find the sunlit leaf area index (L^*) for the entire horizontal leaf canopy by subtracting the fractional sunlit area below the canopy from 1:

$$L^* = 1 - e^{-L_t}. \tag{10.4}$$

To see what Equation 10.4 is telling us, we could find L^* for several values of L_t. These are shown in Figure 10.1. We see that for horizontal leaves, as leaf area index increases beyond about 3, sunlit leaf area approaches its maximum value of 1.

Equation 10.4 could be used along with the photosynthesis and transpiration equations from Chapter 9 to predict canopy photosynthesis and transpiration if we knew wind, temperature, vapor density, and CO_2 concentration distributions within the canopy. This type of model would not be very useful, however, because few canopies meet our assumptions.

We can make the models somewhat more realistic by considering other leaf orientations and inclinations. We again consider canopies with random leaf distribution and orientation, but with either constant or random inclination. The problem can be simplified considerably by recognizing that it reduces to the horizontal leaf problem if we multiply the measured leaf

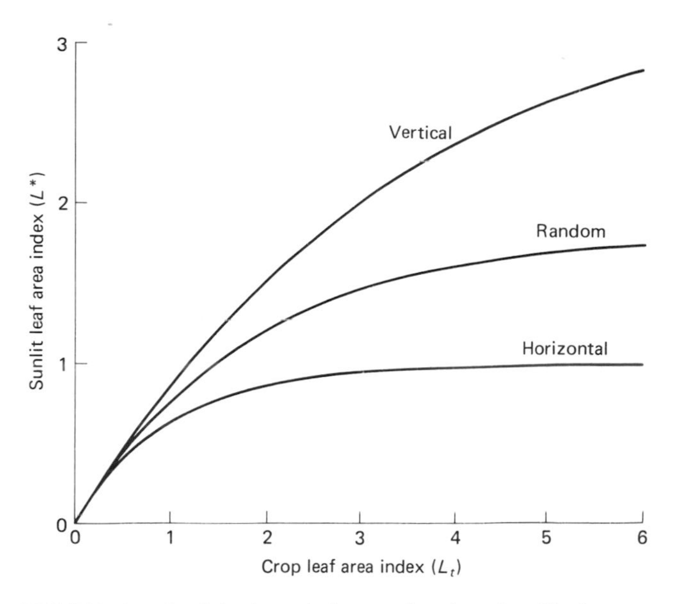

FIGURE 10.1. Sunlit leaf area index as a function of total leaf area index of the crop and canopy structure. The random leaf line approximates an inclined leaf canopy with 60-degree leaf inclination. (Solar elevation angle is 65.5 degrees)

area index by the ratio of leaf shadow area on a horizontal surface to actual leaf area (A_h/A). This is somewhat similar to our approach for finding direct short-wave radiation for animals in Chapter 7, except we are now projecting the leaf area on a horizontal plane rather than a plane perpendicular to the solar beam. If we let $K = A_h/A$, then, for any canopy, the direct short-wave radiation is

$$S_b = S_b^0 \, e^{-KL} \tag{10.5}$$

and

$$L^* = \frac{(1 - e^{-KL_t})}{K} \,. \tag{10.6}$$

These reduce to Equations 10.2 and 10.4 for horizontal leaves since $K = A_h/A = 1$.

Vertical leaf inclination is easily treated if one recognizes that the possible leaf orientations result in an angular distribution of surface area similar to that of a vertical cylinder, half of which is illuminated on the convex side and half on the concave side. This is shown in Figure 10.2. The total surface area of the cylinder walls is $A = \pi dh$ and the area of shadow cast by both halves on a horizontal plane is $A_h = 2dh/\tan\phi$. For vertical inclination we therefore have

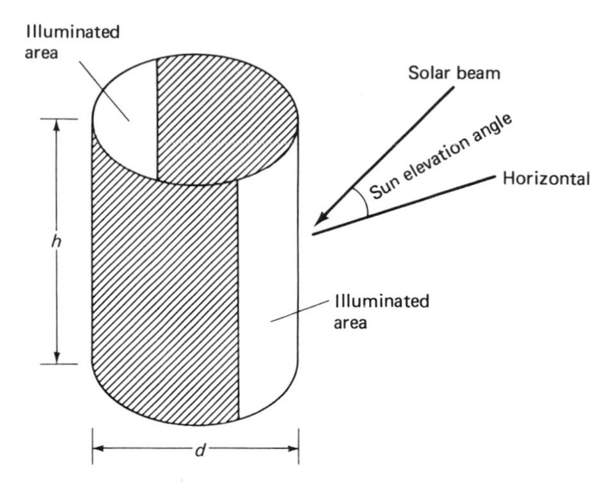

FIGURE 10.2. Cylinder, showing interception of direct radiation equivalent to vertical leaf inclination.

$$K_{\text{vertical}} = \frac{A_h}{A} = \frac{2}{\pi \tan \phi}. \tag{10.7}$$

If the leaf inclination and orientation are random, the distribution of leaf area is similar to the distribution of orientation and inclination for a spherical surface. If we follow a line of reasoning similar to that for vertical leaves ($A = \pi d^2$, $A_h = \pi d^2/2 \sin \phi$), we find

$$K_{\text{random}} = \frac{1}{2 \sin \phi}. \tag{10.8}$$

When leaves have some constant inclination that is neither horizontal nor vertical, they can be thought of as having inclinations and orientations similar to those found on the surface of a cone (Figure 10.3).

If the cone elevation angle is α, then for $\phi > \alpha$, $A_h = \pi d^2/4$ (the area of a circle). The surface area of a cone is $A = \pi dh/2$, where h is the length shown in Figure 10.3, so for $\phi > \alpha$

$$K_{\text{inclined}} = \frac{A_h}{A} = \frac{d}{2h} = \cos \alpha \tag{10.9}$$

since $d = 2h \cos \alpha$. When $\alpha = 0$, this reduces to the horizontal leaf equation.

When $\phi < \alpha$, A_h is the area of the two shadows in Figure 10.3c. The area can be shown to be

$$A_h = \frac{d^2(2 \tan \beta + \pi - 2\beta)}{4}$$

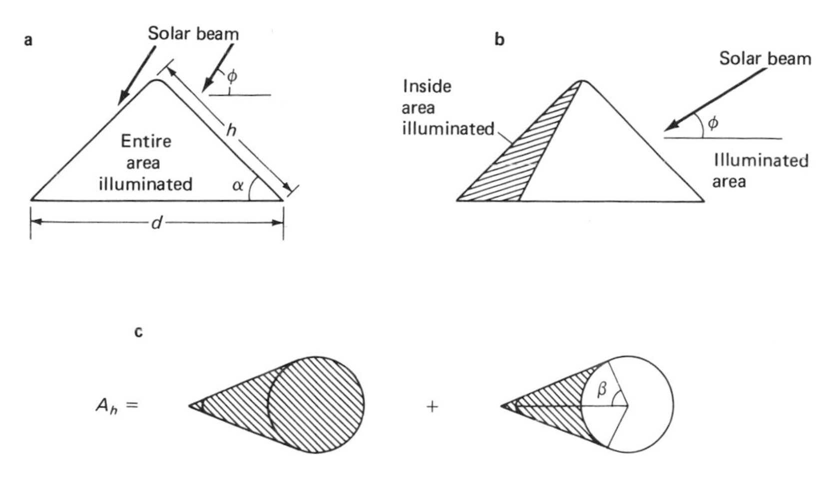

FIGURE 10.3. Cone, showing illuminated areas for solar elevation angles greater than **(a)** and less than **(b)** the cone elevation angle. Shadow areas making up A_h are shown in **(c)**.

where $\cos \beta = \tan \phi / \tan \alpha$ and β is in radians. Thus, for $\phi < \alpha$

$$K_{\text{inclined}} = \frac{A_h}{A} = \left[1 + \frac{2(\tan \beta - \beta)}{\pi}\right] \cos \alpha. \qquad (10.10)$$

Figure 10.1 shows sunlit leaf area index for various canopy structures as a function of total leaf area index for a solar elevation angle of 65.50 (summer solstice at 48° N. latitude). Canopies with inclined leaves obviously have larger sunlit leaf area fractions than horizontal leaf canopies. Since photosynthesis generally light-saturates at light levels well below full sun (Figure 9.1) canopy photosynthesis can be increased by decreasing the flux per unit leaf area and increasing the sunlit leaf area, as is the case with inclined and vertical leaf canopies.

Obviously the leaves in real canopies are generally neither randomly distributed nor all at a given inclination, but these are complications with which we will not concern ourselves here. Our model can be extended to include diffuse short wave from the sky, which, in some areas, may be more significant to photosynthesis than direct shortwave. The leaf area index irradiated by diffuse short wave can be found by integrating Equation 10.6 over all values of ϕ. For horizontal leaves, L^* is independent of ϕ, so the diffuse L^* is just $\pi/2$ times the sunlit L^*. Diffuse L^* values for horizontal leaves are shown in Figure 10.4. At low leaf area index, essentially all leaves receive diffuse short wave, but when L reaches 4, diffuse L^* is close to its upper limt of $\pi/2$. Integration of Equation 10.6 is more difficult for other canopy structures. Figure 10.4 shows diffuse L^* as a function of L_t for vertical, inclined and random

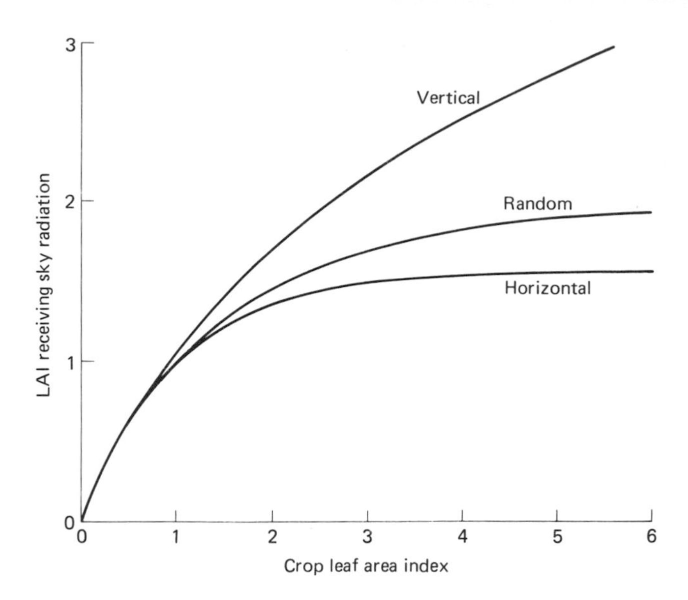

FIGURE 10.4. Leaf area per unit ground area that receives diffuse sky radiation as a function of total leaf area index of the crop for three canopy structures. Inclined leaves at 30 degrees follow the horizontal leaf line. Inclined leaves at 60 degrees follow the random leaf line.

leaves. Again, a vertical leaf canopy has a higher diffuse L^* than a horizontal canopy. Additional details on canopy models can be found in Norman [10.11].

We can illustrate the use of the canopy models by comparing predicted photosynthesis of the model canopies using photosynthesis equations from Chapter 9. We will assume that photosynthesis is limited only by light, that we have only direct short wave, and that the photosynthetically active radiation above the canopy is half the value given by Equation 5.7 (using Equations 5.9 and 5.10 to obtain S_p and m). The atmospheric transmission coefficient, a in Equation 5.9 is taken as 0.84 for this simulation. The elevation angle ϕ is related to location and time of day using Equation 5.8.

A proper calculation of photosynthesis would require that we find the irradiance for each sunlit leaf, compute the photosynthesis for that leaf, and add up the contributions from all leaves in the canopy. This would be very difficult. We will resort to the approximate approach of computing the average irradiance on a sunlit leaf and using Equation 10.6 to find the leaf area that receives this radiation. The average irradiance for a sunlit leaf is the light absorbed by the canopy divided by the total sunlit area. Equation 10.5 gives the fractional sunlit area below the canopy as $S_b/S_b^0 = e^{-KL_t}$. The fractional area of shadow is $1 - e^{-KL_t}$, so the absorbed radiation is $PAR^0 (1 - e^{-KL_t})$. The average irradiance on a leaf is therefore (using Equation 10.6)

$$PAR = \frac{PAR^0 (1 - e^{-KL_t})}{\frac{1 - e^{-KL_t}}{K}} = K\, PAR^0. \tag{10.11}$$

Using these assumptions, values from Table 9.1, Equations 9.3 and 9.4, and the equations for K just developed, we will find the photosynthesis per unit ground surface area for various distributions of inclination. The declination and latitude for this simulation will be $\delta = 21.9°$ and $\lambda = 48°$ (June in Pullman). Results of the calculations are shown in Figure 10.5 for three values of L_t. Table 10.1 shows daily gross photosynthesis for each of the curves in Figure 10.5.

For production estimates we need net rather than gross photosynthesis. We must obtain estimates of respiration for the canopy, and substract these from the values shown in Table 10.1. McCree [10.5] divided respiration into maintenance and growth components. The maintenance component was assumed proportional to total canopy mass, and the growth component to gross photosynthesis. The equation

$$R = aP + bW \tag{10.12}$$

was used to describe daily respiration, where P is gross photosynthesis, W is the standing dry matter mass expressed as CO_2 equivalent (44/30 × dry mass), and a and b are constants. For white clover at 20°C, McCree found $a = 0.25$ and $b = 0.015$ day^{-1}. Biscoe *et al.* [10.1] extended the treatment to include temperature dependence of respiration, but confirm McCree's constants for barley under field conditions. We will use McCree's equation to find net photosynthesis for the values in Table 10.1. Equation 10.12, with McCree's constants, says that about one-fourth of the photosynthate produced is respired for growth, and that 1.5 percent of the crop dry matter is used each day for maintenance.

To find the rate of maintenance respiration, we need to have a relationship between leaf area index and standing dry matter. The leaf area ratio is the ratio of total leaf area to total dry mass of a canopy. A typical leaf area ratio would be 10–15 m²/kg [10.7]. Converting 10 m²/kg to CO_2 equivalent mass gives

$$bW = 2.2\, L_t \text{ (g m}^{-2}\text{ day}^{-1}) \tag{10.13}$$

Table 10.1 shows net photosynthesis for various combinations of canopy structure and leaf area index. In Figure 10.6, net

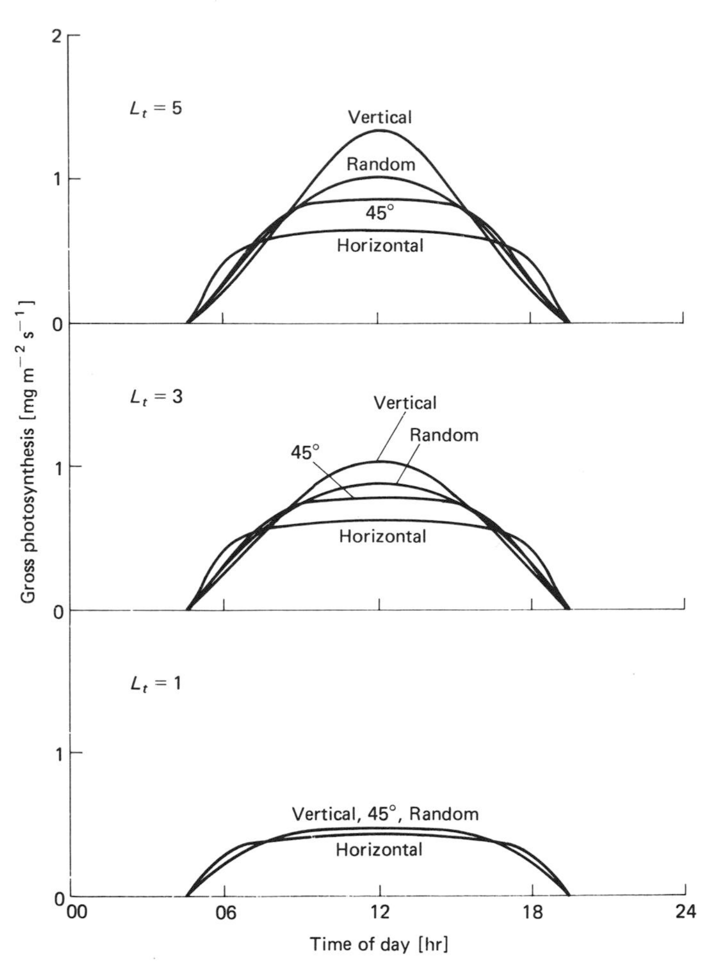

FIGURE 10.5. Gross photosynthetic rate in model canopies having various leaf areas and leaf distributions as a function of time of day. Simulations are for 48 °N. latitude (Pullman) on June 1.

Table 10.1 Average photosynthesis per unit ground area as a function of leaf area index and leaf inclination

Leaf inclination	Gross photosynthesis (g m^{-2}day^{-1})			Net photosynthesis (g m^{-2}day^{-1})		
	$L_t = 1$	$L_t = 3$	$L_t = 5$	$L_t = 1$	$L_t = 3$	$L_t = 5$
Horizontal ($\alpha = 0$)	18.6	27.9	29.2	11.8	14.3	10.9
$\alpha = 45°$	19.4	31.4	34.0	12.4	17.0	14.5
Vertical ($\alpha = 90$)	19.1	34.1	39.8	12.1	19.0	18.9
Random	19.4	32.3	35.6	12.4	17.6	15.7

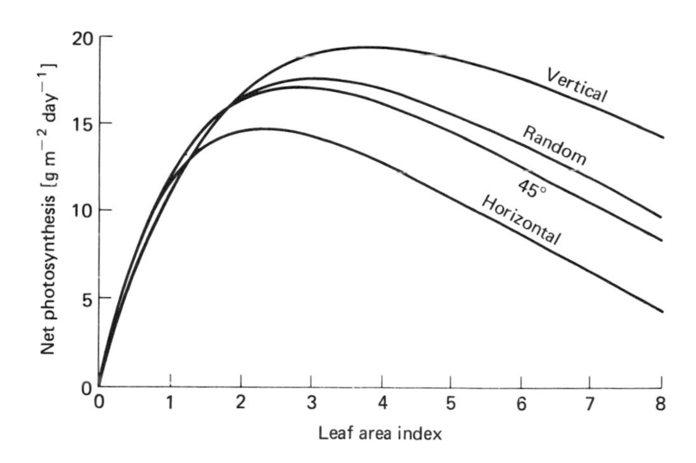

FIGURE 10.6. Canopy net photosynthesis for model canopies as a function of leaf area index and canopy structure.

photosynthesis is plotted as a function of leaf area index. We note that there is an optimum leaf area index for production, and that the optimum value of L_t depends on canopy structure. If this simulation is reasonably correct, and in the absence of other constraints, we would expect natural canopies to maintain a leaf area index of 2 to 4 depending on structure. We would also expect maximum production from agricultural crops that have erect leaf canopies with $L_t \simeq 4$. These predictions are borne out reasonably well by observation. The usefulness of such predictions to crop breeding research is quite obvious when one realizes that, with high fertility and intensive management, L_t values as high as 12 are sometimes attained for part of a growing season. This is well past the optimum, and much of the overproduction of leaf area is wasted as leaves senesce. Such dense canopies also create favorable microenvironments for disease organisms and insects. Development of genetic limitations to leaf area index of intensively managed crops has been one of the triumphs of the "green revolution."

Evapotranspiration

An understanding of evapotranspiration and the ability to estimate it are important to disciplines such as irrigation system design, crop and forest production, watershed management, and land and water resource planning. Equations of the type we will develop here have been used to predict water loss from bare soil, reservoirs, lakes, crops, and forests. Several good reviews of this subject are available and can be consulted for further details [10.3, 10.13]. We will develop equations from the principles we have already discussed, starting with equations describing instantaneous flux from a suface, and then making appropriate assumptions and simplifications to obtain averages.

It is possible to estimate evapotranspiration using several different approaches. We could consider evaporation from individual leaves or surface elements and sum these to find total evaporation. If we were to do this we would need to know windspeeds and vapor densities within the canopy or near the evaporating surface in detail. Rather than attempt this, we will look at the exchange surface as an entity that has a temperature, a vapor density, and a diffusion resistance, and use the turbulent transport equations from Chapter 4 to find the rate of water loss. The problem then becomes similar to the energy budget problem for individual leaves or animals.

The Energy Budget

From our past discussions we know that evaporation is an important part of the energy budget. It should be possible to infer evaporation by measuring all of the other terms in the energy budget and taking E as what is left. The energy budget for a crop, soil, or water surface is

$$R_n - G - H - \lambda E = 0 \tag{10.14}$$

where R_n is the net radiation for the surface, G the rate of heat storage in the soil or water, H is the sensible heat loss, and λ is the latent heat of vaporization. We can estimate or measure R_n and G, and H and E depend on eddy transport. We can simplify the problem somewhat by taking the ratio, $\beta = H/\lambda E$, called the Bowen ratio. Substituting for H in Equation 10.14 and solving for λE, we obtain

$$\lambda E = \frac{R_n - G}{1 + \beta}. \tag{10.15}$$

If we can evaluate β, Equation 10.15 should give us estimates of λE that are relatively insensitive to atmospheric transport properties, at least when $\beta < 1$. From Equations 6.1 and 6.2 we can write

$$\beta = \frac{H}{\lambda E} = \frac{\rho c_p (T_1 - T_2)}{\lambda(\rho_{v1} - \rho_{v2})} \frac{r_v}{r_H} \tag{10.16}$$

where the subscripts 1 and 2 indicate values of the variables at heights z_1 and z_2, and r_v and r_H are the vapor and heat transfer resistances between heights z_1 and z_2. Thus if temperature and vapor density are measured at two heights, β can be calculated (assuming $r_v = r_H$). A typical value for β over well-watered short grass or a wet soil surface is around 0.2. As the surface dries, β becomes larger and approaches infinity for a dry surface. The Bowen ratio may become negative in arid areas as a result of advection (horizontal heat transport from surrounding areas). Heat is extracted from the air that moves over the crop to evaporate water from the leaves. This results in lower temperatures near the crop than in the air above the crop making $T_1 - T_2$ negative.

In the atmosphere above the surface, the assumption that $r_v = r_H$ is probably quite good. We could, however, choose T_1 to be the surface temperature and ρ_{v1} to be the corresponding saturation vapor density. If the surface vapor resistance r_{vc} is zero (surface saturated), the assumption that $r_v = r_H$ is still good, so with a measurement of surface and air temperature and air vapor density, we could find the Bowen ratio. This is illustrated by the following example.

Example. Assume $\rho_{va} = 6$ g/m³, $T_a = 20°$C, $T_s = 25°$C, and $R_n = 500$ W/m² for a wet, bare soil surface. What is E?

Solution. To solve Equation 10.15 we need to know G and β. A reasonable approximation for daytime conditions is $G = 0.1\ R_n$. From Equation 10.16 with $r_v = r_H$, we obtain

$$\beta = \frac{1200 \text{ J m}^{-3}\text{C}^{-1}(25\text{-}20)\text{C}}{2430 \text{ J/g } (23\text{-}6)\text{g/m}^{-3}} = 0.15.$$

From Equation 10.15,

$$E = \frac{0.9 \times 500 \text{ W/m}^2}{2430 \text{ J g}^{-1}(1 + 0.15)} = 0.16 \text{ g m}^{-2}\text{s}^{-1}.$$

Use of Equation 10.15 is limited by the data requirement, particularly for determining the Bowen ratio. The data requirement is relaxed somewhat if we use Penman's method to eliminate surface temperature from the equation, as we did in Chapter 9. The Penman equation (Equation 9.10) for a crop or soil surface is

$$\lambda E = \frac{s(R_n - G) + \dfrac{\rho c_p(\rho'_{va} - \rho_{va})}{r_H}}{\gamma^* + s} \tag{10.17}$$

where $\gamma^* = \gamma r_v/r_H$. When we derived this equation we assumed that the evaporating surface temperature was the same as the surface temperature for convective exchange. This assumption was obviously valid for a single leaf, but may be open to question when the equation is applied to a soil surface or a crop where the primary surface for heat exchange may be quite different from the primary surface for vapor exchange. For dense, well-watered crop canopies, the heat exchange surface is the vapor exchange surface, and the surface vapor resistance can usually be considered negligible compared to boundary layer resistance, so $r_v = r_H$. We will use the term "potential evapotranspiration" to describe this set of conditions.

Potential Evapotranspiration

Equation 10.17, with $r_v = r_H$, can be used to find potential evapotranspiration. The boundary layer resistance for a surface can be found by combining Equations 4.9, 4.8, and 6.1 to give

$$r_H = \frac{\ln\left(\frac{z - d + z_H}{z_H}\right)\ln\left(\frac{z - d + z_M}{z_M}\right)}{k^2\,\bar{u}} \tag{10.18}$$

where $\bar{u}$ is the mean windspeed measured at height z. If the roughness parameters for heat and vapor are the same, then $r_H = r_v$. Substitution of Equation 10.18 into Equation 10.17 with $r_v = r_H$ gives

$$\lambda E_p = \frac{s}{s + \gamma}(R_n - G) + \frac{\gamma}{s + \gamma}\, f(u)(\rho'_{va} - \rho_{va}) \tag{10.19}$$

where $f(u) = \lambda/r_H$. Equation 10.19 is in a form that contains easily estimated or measured climatological and crop variables.

Table A.3 gives values for s, $s/(s + \gamma)$, and ρ'_{va} as functions of temperature. Heat storage rate (G) can be measured or estimated. A reasonable daytime approximation for crops is $G = 0.1\, R_n$. For daily averages, G can be taken as zero. The windspeed at height z and the air vapor density are usually measured. The saturation vapor density is found directly from air temperature. The air vapor density remains fairly constant over periods of a day unless there is an airmass change or a strong source of water vapor such as a sea breeze. Thus a single daily measurement of ρ_{va} will usually suffice. If no vapor density data are available at all, one can estimate ρ_{va} by assuming that the air cools to about dewpoint temperature on clear, calm nights. Minimum temperatures for these nights can be used to provide approximate values for ρ_{va}. Values for z_H, z_M, and d can be obtained from empirical expressions given in Chapter 4.

When the Penman equation is used for daily evapotranspiration estimates, an empirical wind function of the form $f(u) = a + b\bar{u}$ is employed. When wind is the daily average in m/s at $z = 2$ m, $f(u)$ can be taken as $5.3(1 + \bar{u})$. For a grass surface with $z_M = 1$ cm and $z_H = 0.2$ cm, the wind function portion of Equation 10.19 is $f(u) = 11\,\bar{u}$ when $\bar{u}$ is measured at $z = 2$ m.

Thus the empirical function has weaker windspeed dependence, but the theoretical function goes to zero at $\bar{u} = 0$. At $\bar{u} = 0.6$ m/s they are equal. The difference may result from the fact that the empirical function is for daily mean values of λE_p while the theoretical function is for instantaneous measurements.

Net radiation is the last term we need to evaluate to estimate λE_p. A direct measurement with a net radiometer is best, but such measurements are generally not readily available. The next best choice is a direct measurement of solar radiation and estimates of net long-wave radiation and albedo. If solar radiation measurements are not available, various methods can be used for making estimates. Jensen and Haise [10.4] and Jensen [10.3] give good summaries of these methods.

The net long-wave radiation can be estimated from

$$L_{\text{net}} = (\epsilon_A - \epsilon_s)\, \sigma T_a^4 \tag{10.20}$$

if one assumes that the surface temperature equals air temperature. Table A.3 or Equation 5.12 or 5.13 can be used to find clear sky ϵ_A. We can assume ϵ_s is 0.97.

To give some idea of the predictive ability of Equation 10.19, Figure 10.7 shows calculated and measured E_p for an alfalfa crop over a 24-hr period. The agreement is quite good.

Monthly averages of calculated E_p, compared to measured values, are shown in Figure 10.8. These were computed using monthly average R_n, T_a, $\bar{u}$, etc., and the empirical wind func-

FIGURE 10.7. Comparison of calculated and measured evapotranspiration from well-watered alfalfa at Tempe, Arizona. (From van Bavel [10.4].)

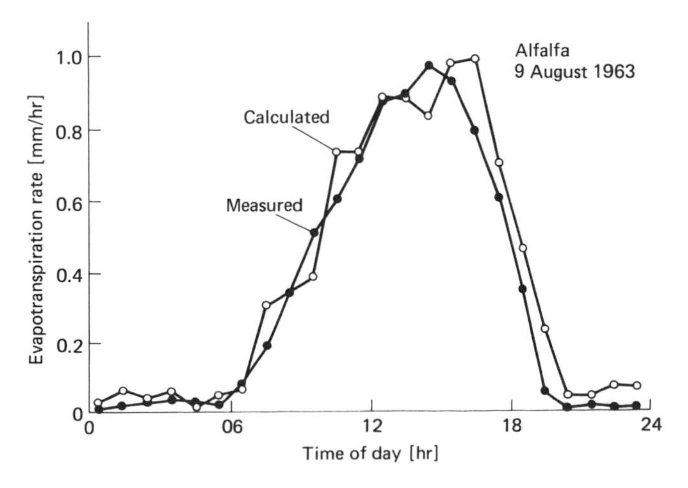

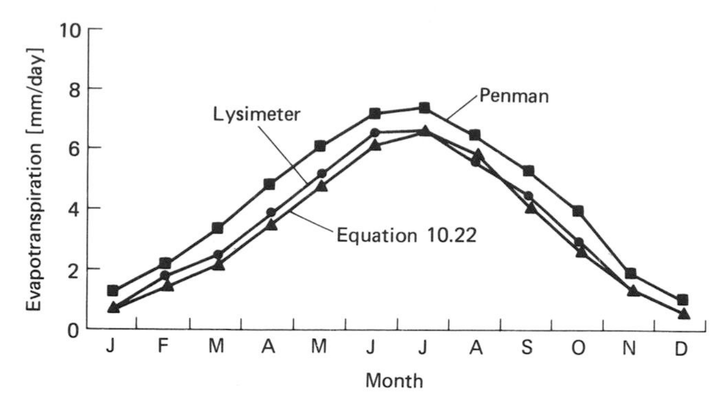

FIGURE 10.8. Average monthly evapotranspiration from a short grass surface measured with a lysimeter at Davis, California, and calculated using the Penman equation and Equation 10.22. (After Jensen [10.3].)

tion. Agreement between predicted and measured E_p is good at Davis. At other locations it is often much poorer because of advection effects. Also, the long averaging times mask large short-term differences between measured and calculated values.

The data requirements for Equation 10.19 are often still excessive, and further simplification is possible if we assume that λE_p depends primarily on energy supply to the evaporating surface. When the air is saturated, Equation 10.18 becomes

$$\lambda E_p = \frac{s}{s + \gamma} (R_n - G). \tag{10.21}$$

Priestly and Taylor [10.12] multiplied this equation by a constant, α (=1.3), to estimate E_p from climatic data. This can be simplified even more by assuming that $G \simeq 0$ when averaged over several days and R_n is proportional to solar irradiance, S_t and T_a. The equation is then written as

$$\lambda E_p = a(T_a + b)S_t \tag{10.22}$$

where a and b are empirically determined. Typical values are $a = 0.025°C^{-1}$ and $b = 3°C$ when S_t is in W/m² [10.3]. Predictions by this equation are also shown in Figure 10.8. The fact that Equation 10.22 agrees better with measured E_p than the Penman equation is probably fortuitous.

Example. What is E_p for a forest with average $S_t = 300$ W/m² and $T_a = 20°C$?

$$\lambda E_p = 0.025°C^{-1}(20 + 3)\ C \times 300\ W/m^2 = 173\ W/m^2$$

$$E_p = \frac{173 \text{ J s}^{-1}\text{m}^{-2}}{2430 \text{ J g}^{-1}} = 0.071 \text{ g m}^{-2}\text{s}^{-1}$$

or $0.071 \text{ g m}^{-2}\text{s}^{-1} \times 8.64 \times 10^4 \text{ s/day} \times 10^{-6}\text{m}^3/\text{g}$ $= 6.1$ mm/day.

Nonpotential Evapotranspiration

The concept of potential evapotranspiration (ET) is useful because it allows us to estimate the rate of water loss from a surface when properties of the surface (other than surface roughness) do not influence rate of water loss. We recognize, however, that few surfaces, other than open water, will evaporate at potential rates all of the time, and most soil and vegetated sufaces will evaporate at potential rates for only a small fraction of the time. Prediction of water loss under nonpotential conditions may therefore be more important to the water budgets of many surfaces than potential ET estimates.

Reduction in ET below the potential rate is the result of stomatal closure or drying at a soil surface. The increase in surface resistance caused by these processes is generally the result of an insufficient rate of water supply to the evaporating surface. Under these conditions, the rate of water supply to the surface determines the rate of evaporation. The resistance is adjusted by the system to make the water loss rate equal to the water supply rate. Because of this, the most fruitful approach to estimation of nonpotential ET is usually to consider rate of supply rather than rate of loss. This has been successfully accomplished in computer simulations of soil drying and water extraction by plants [10.10]. Simpler, more empirical approaches have been used extensively [10.9] and work well for many purposes. Numerous attempts have been made to express nonpotential ET as some fraction of potential ET. If the soil or soil-plant system (rather than atmospheric factors) is controlling water loss then such attempts are obviously futile because ET is not functionally related to potential ET.

We can make some interesting observations about nonpotential ET under specific conditions. The results will not apply generally, but may give an indication of possible approaches for solving specific problems. McNaughton and Black [10.6] analyzed ET for a conifer forest using an energy balance approach. They show that vapor concentration and temperature within the canopy are essentially the same as above the canopy because turbulent mixing is so intense in a forest. We already know that conifer leaves remain close to air temperature because of their small size. Also, because of their small size and high stomatal diffusion resistance (typically 500 to 1000 s/m), boundary layer resistance is negligible compared to stomatal resistance. The rate of water loss from the forest is therefore completely determined by Equation 6.2, with ρ'_{vs} the saturation value at air temperature and $r_v = r_{vc}$. The canopy diffusion resistance for this calculation is just the equivalent parallel resistance of all of the leaf layers. If the stomatal diffusion resistance for all layers is the same, then $r_{vc} = r_{vs}/L_t$. The transpiration from a conifer canopy can therefore be estimated from a single measurement of air temperature and vapor density (at essentially any height) and an estimate of stomatal diffusion resistance and leaf area index. The absence of good stomatal diffusion resistance data for conifers has made application of such an approach impractical in the past, but modern equipment for making these measurements will allow a more thorough exploitation of the method. The specific characteristics of the canopy and leaves make it possible to simplify the ET calculation for this specific case. This simplification would not work in most situations, however.

We can learn about possible plant reactions that control water loss under water stress conditions by determining the increase in surface diffusion resistance necessary to reduce ET a given amount. For a reasonably dense canopy, we can probably safely assume that heat and vapor are exchanged by the same surface, even when the crop is water-stressed. As the diffusion resistance of the crop increases, heat dissipation will change from primarily latent to primarily sensible and radiant loss. Since the partitioning of energy changes, an analysis based on the Penman equation is not very useful because the net radiation term will change with the condition of the surface. If we use the Penman transformation to linearize the vapor density term, and Equation 7.16 to linearize the net long-wave radiation, we obtain an equation for latent heat loss which, except for roughness and crop diffusion resistance, is completely environment-dependent:

$$\lambda E = \left(\frac{s}{s + \gamma^*}\right)\left[R_{\text{abs}} - G - \epsilon\sigma T_a^4 + \frac{\rho c_p(\rho'_{va} - \rho_{va})}{s r_e}\right] \tag{10.23}$$

where $\gamma^* = \gamma r_v/r_e$, $r_v = r_{va} + r_{vc}$, and $1/r_e = 1/r_H + 1/r_r$. The main advantage of this equation over the conventional Penman equation is that surface temperature is eliminated from the net radiation term as well as the other terms. When r_{vc} is zero, Equation 10.23 gives potential ET. Now, for constant radiant energy input and constant wind, we can find the ratio of ET to potential ET as a function of crop diffusion resistance. Dividing out common terms, the ratio is

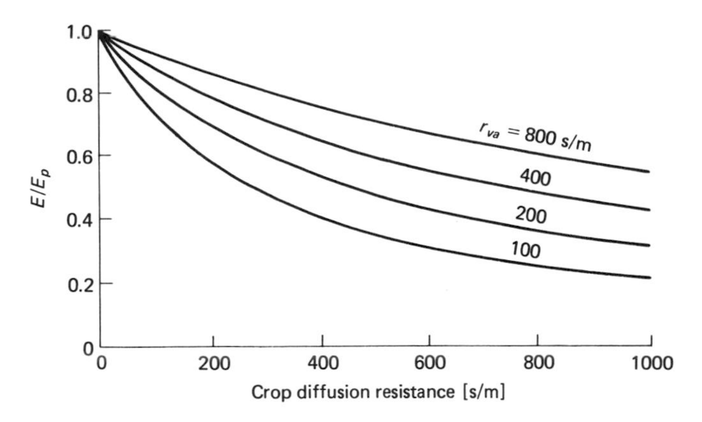

FIGURE 10.9. Ratio of actual to potential evapotranspiration as a function of crop boundary layer diffusion resistance.

$$\frac{E}{E_p} = \frac{s + \gamma \dfrac{r_{va}}{r_e}}{s + \gamma \dfrac{r_{va} + r_{vc}}{r_e}}.$$

Figure 10.9 shows the ratio of ET to potential ET as a function of boundary layer and crop diffusion resistance at $T_a = 25°C$. Note that crop diffusion resistances must be around 500 s/m before transpiration is reduced to even half the potential rate for typical boundary layer resistances. If we assume that only direct or diffuse sunlit leaves have open stomates, then the maximum leaf area index that can contribute to transpiration is around 2 (Figure 10.4). The crop diffusion resistance can be taken as roughly r_{vs}/L^*, or about half the diffusion resistance of individual leaves. Therefore the leaf diffusion resistance must reach levels of at least 1000 s/m to effectively control water loss. Photosynthesis is markedly reduced at such high diffusion resistances, and, with high respiratory losses from the high leaf temperatures, it would be difficult to maintain a positive carbon balance in the leaves. Reduction in leaf area allows the plant to maintain low leaf diffusion resistances while, at the same time, maintaining high crop resistances to reduce ET. Reduction in leaf area is a typical response to drought when the drought is imposed over a sufficiently long period of time to allow the plant to adjust to the stress.

References

10.1 Biscoe, P. V., R. K. Scott, and J. L. Monteith (1975) Barley and its environment III. Carbon budget of the stand. *J. Appl. Ecol.* *12*:269–293.

10.2 Fuchs, M. (1972) The control of the radiation climate of plant communities. *Optimizing the Soil Physical Environment Toward Greater Crop Yields* (D. Hillel, ed.). New York: Academic Press.

10.3 Jensen, M. E. (ed.) (1973) *Consumptive Use of Water and Irrigation Water Requirements*. New York: American Society of Civil Engineers.

10.4 Jensen, M. E. and H. R. Haise (1963) Estimating evapotranspiration from solar radiation. *J. Irrig. Drain. Div., Am. Soc. Civ. Engr. 89*:15–41.

10.5 McCree, K. (1970) An equation for the rate of respiration of white clover plants grown under controlled conditions. *Prediction and Measurement of Photosynthetic Productivity* (I. Setlick, ed.). Wageningen: Centre for Agricultural Publishing and Documentation.

10.6 McNaughton, K. G. and T. A. Black (1973) A study of evapotranspiration from a Douglas fir forest using the energy balance approach. *Water Resour. Res. 9*:1579–1590.

10.7 Milthorpe, F. L. and J. Moorby (1974) *An Introduction to Crop Physiology*. New York: Cambridge University Press.

10.8 Monteith, J. L. (1973) *Principles of Environmental Physics*. New York: American Elsevier.

10.9 Munn, R. E. (1970) *Biometeorological Methods*. New York: Academic Press.

10.10 Nimah, M. N. and R. J. Hanks (1973) Model for estimating soil water, plant, and atmospheric interrelations: I. description and sensitivity. *Soil Sci. Soc. Am. Proc. 37*:522–527.

10.11 Norman, J. M. (1975) Radiative transfer in vegetation. *Heat and Mass Transfer in the Biosphere*. I. *Transfer Processes in Plant Environment* (D. A. deVries and N. H. Afgan, eds.). New York: John Wiley.

10.12 Priestly, C. H. B. and R. J. Taylor (1972) On the assessment of surface heat flux and evaporation using large-scale parameters. *Mon. Weather Rev. 100*:81–92.

10.13 Tanner, C. B. (1968) Evaporation of water from plants and soils. *Water Stress and Plant Growth,* Vol. 1. (T. Kozlowski, ed.). New York: Academic Press.

10.14 van Bavel, C. H. M. (1966) Potential evaporation: the combination concept and its experimental verification. *Water Resour. Res. 2*:455–467.

Problems

10.1 Find the sunlit leaf area and the sunlit ground area for a horizontal leaf canopy with a leaf area index of 3.5.

10.2 Compare canopy photosynthesis for vertical and horizontal leaf canopies at noon on July 1 in Pullman.

10.3 Compare E_p estimates from Equation 10.19, Equation 10.19 with the empirical wind function $f(u) = 5.3(1 + u)$, Equation

10.21, Equation 10.22, and Equation 10.23 for the following data: S_t = 239 W/m^2, albedo = 0.25, $\bar{u}$ = 1.61 m/s, $\overline{T}$ = 5.8°C, ρ_{va} = 4.2 g/m^3. Assume $T_s = T_a$ for the net long-wave calculation.

Appendix
Reconciliation of Terminology

Heat and mass transport are expressed in terms of transport resistances in this book. Advantages of this approach are discussed in Chapter 1. Several alternative methods of expressing transport have been used extensively, particularly in the animal physiology literature. This section is provided as an aid for those who are more familiar with conductance expressions.

The rate of convective heat transfer per unit area of surface can be expressed as

$$H = h_c(T_s - T_a) \tag{A.1}$$

where h_c is the area specific convection coefficient and T_s and T_a are surface and ambient temperatures. Comparison of Equation A.1 with Equation 6.1 gives the relationship between the convection coefficient and the heat transfer resistance:

$$h_c = \frac{\rho c_p}{r_H}. \tag{A.2}$$

Units of h_c in the International System are W m^{-2} K^{-1}. Table A.4 shows conversions from other units to SI units.

Latent heat transport is sometimes expressed as

$$\lambda E = \frac{h_E(\rho_{vs} - \rho_{va})}{\gamma} \tag{A.3}$$

where ρ_{vs} and ρ_{va} are the vapor densities in the air and at the exchange surface, λ is the latent heat of vaporization, and γ is the psychrometric constant ($= \rho c_p/\lambda$).

Comparison of Equation A.3 with Equation 6.2 gives

$$h_E = \frac{\rho c_p}{r_v}.$$

The rate of radiant energy exchange between an animal surface and a blackbody enclosure is often expressed in a linearized form as

$$\epsilon\sigma(T_s^4 - T_r^4) \simeq h_r(T_s - T_r) \qquad \text{(A.4)}$$

where T_r is the mean radiant temperature of the surroundings and $h_r = 4\epsilon\sigma T^3$, T being the average of T_s and T_r. The definition of the radiative resistance, r_r (Equation 7.10) indicates that

$$h_r = \frac{\rho c_p}{r_r}\,. \qquad \text{(A.5)}$$

The radiative and convective conductances are sometimes combined to form a ''dry heat transfer coefficient,'' h_e. The relationship between this coefficient and the parallel equivalent resistance for convection and radiation, r_e (Equation 7.10) is

$$h_e = \frac{\rho c_p}{r_e}\,. \qquad \text{(A.6)}$$

Conductances have sometimes been defined on a mass- rather than area-specific basis for use in correlating data. In other studies a transfer coefficient has been obtained by dividing metabolic heat production by the difference between body and air temperature. Conversion of these conductances to values that can be used for animal energy budget studies is sometimes difficult and involves a number of assumptions. Some methods for conversion are given in Calder and King [A.1] and Robinson *et al.* [A.3].

The blackbody equivalent temperature, T_e, used in this book (Equation 7.17) is identical to the ''effective temperature for the environment'' used by Monteith [A.2], and is a more general form of the ''operative temperature'' proposed by Winslow *et al.* [A.4] and used in many human environmental physiology studies. Operative temperature is expressed as either

$$T_{op} = \frac{h_r T_r + h_c T_a}{h_r + h_c} \qquad \text{(A.7)}$$

or

$$T_{op} = T_a + \frac{h_r(T_r - T_a)}{h_e} \qquad \text{(A.8)}$$

where T_r is the mean radiant temperature of the surroundings, and energy derived from short-wave sources is considered negligible. Equations A.7 and A.8 are equivalent expressions. The equivalence of Equations A.8 and 7.17 is established by noting that, for a blackbody enclosure, $R_{abs} = \epsilon\sigma T_r^4$. The expression $R_{abs} - \epsilon\sigma T_a^4$ can then be approximated as $h_r(T_r - T_a)$ (Equation A.4). Substitution for r_e from Equation A.6 yields Equation A.8.

In this book the terminology of soil and plant scientists has been adopted for expressing the energy status of water in the living organism and its environment. Animal physiologists express the energy status of water in terms of a concentration of an ideal solution. Units are typically osmoles per kilogram where one osmole is one mole of an ideal solute. The relationship between water potential and osmolarity is given by the van't Hoff equation (Equation 3.8)

$$C = \frac{-\Psi}{RT} \tag{A.9}$$

where C is the osmolarity (osmoles/kg), Ψ is the water potential (J/kg), R is the gas constant (8.31 J K^{-1} mol^{-1}), and T is the Kelvin temperature. At 25° C, 1 J/kg is equal to 0.4 milliosmoles per kilogram.

Table A.1 Temperature-dependent properties of air at 100 k Pa pressure

T[°C]	ρ[kg/m³]	υ[mm²/s]	D_H [mm²/s]	D_v[mm²/s]	D_c[mm²/s]	D_o[mm²/s]
0	1.292	13.3	18.9	21.2	12.9	17.7
5	1.269	13.7	19.5	22.0	13.3	18.3
10	1.246	14.2	20.2	22.7	13.8	19.0
15	1.225	14.6	20.8	23.4	14.2	19.6
20	1.204	15.1	21.5	24.2	14.7	20.2
25	1.183	15.5	22.2	24.9	15.1	20.8
30	1.164	16.0	22.8	25.7	15.6	21.5
35	1.146	16.4	23.5	26.4	16.0	22.1
40	1.128	16.9	24.2	27.2	16.5	22.7
45	1.110	17.4	24.9	28.0	17.0	23.4

Specific heat of air is c_p = 1.01 kJ kg^{-1}°C^{-1}.

Table A.2 Properties of water

T[°C]	ρ[Mg/m³]	λ[MJ/kg]	υ[mm²/s]	D_H[mm²/s[	D_o[mm²/s]	D_c[mm²/s]
0	0.99987	2.50	1.79	0.134		
4	1.00000	2.49	1.57	0.136		
10	0.99973	2.48	1.31	0.140		
20	0.99823	2.45	1.01	0.144	2×10^{-3}	2×10^{-3}
30	0.99568	2.43	0.80	0.148		
40	0.99225	2.41	0.66	0.151		
50	0.98807	2.38	0.56	0.154		

Specific heat of water is c = 4.19 kJ kg^{-1}°C^{-1}.

Table A.3 Temperature dependence of blackbody emittance, radiative resistance, clear-sky emissivity, saturation vapor density, slope of the saturation vapor density function, and two derived parameters

Temperature (°C)	Blackbody emittance (W/m²)	r_r (s/m)	Clear-sky emissivity	Saturation vapor density (g/m³)	s (g m^{-3} K^{-1})	$\frac{s}{s+\gamma}$	$\frac{\gamma}{s+\gamma}$
−5	293	274	0.70	3.41	0.25	0.32	0.68
−4	298	271	0.70	3.66	0.26	0.33	0.67
−3	302	268	0.71	3.93	0.28	0.35	0.65
−2	306	265	0.71	4.22	0.30	0.36	0.64
−1	311	262	0.72	4.52	0.32	0.38	0.62
0	316	260	0.72	4.85	0.33	0.39	0.61
1	320	257	0.73	5.19	0.36	0.41	0.59
2	325	254	0.73	5.56	0.38	0.42	0.58
3	330	251	0.74	5.95	0.40	0.44	0.56
4	335	249	0.74	6.36	0.42	0.45	0.55
5	339	246	0.75	6.80	0.45	0.47	0.53
6	344	243	0.75	7.26	0.48	0.48	0.52
7	349	241	0.76	7.75	0.51	0.50	0.50
8	354	238	0.76	8.27	0.53	0.51	0.49
9	359	236	0.77	8.82	0.57	0.53	0.47
10	364	233	0.77	9.40	0.60	0.54	0.46
11	370	231	0.78	10.02	0.63	0.55	0.45
12	375	228	0.78	10.66	0.67	0.57	0.43
13	380	226	0.79	11.35	0.70	0.58	0.42
14	385	223	0.79	12.07	0.74	0.60	0.40
15	391	221	0.80	12.83	0.78	0.61	0.39
16	396	219	0.80	13.64	0.83	0.62	0.38
17	402	217	0.81	14.49	0.87	0.63	0.37
18	407	214	0.81	15.38	0.92	0.65	0.35
19	413	212	0.82	16.32	0.96	0.66	0.34
20	419	210	0.82	17.31	1.01	0.67	0.33
21	424	208	0.83	18.35	1.07	0.68	0.32
22	430	206	0.83	19.44	1.12	0.69	0.31
23	436	204	0.84	20.59	1.18	0.70	0.30
24	442	202	0.84	21.80	1.24	0.72	0.28
25	448	200	0.85	23.06	1.30	0.73	0.27
26	454	198	0.85	24.39	1.36	0.74	0.26
27	460	196	0.86	25.79	1.43	0.75	0.25
28	466	194	0.86	27.25	1.50	0.75	0.25
29	473	192	0.87	28.79	1.57	0.76	0.24
30	479	190	0.87	30.40	1.65	0.77	0.23
31	485	188	0.88	32.08	1.72	0.78	0.22
32	492	186	0.88	33.85	1.81	0.79	0.21
33	498	184	0.89	35.69	1.89	0.80	0.20
34	505	183	0.89	37.63	1.98	0.80	0.20
35	511	181	0.90	39.65	2.07	0.81	0.19
36	518	179	0.90	41.76	2.16	0.82	0.18
37	525	177	0.91	43.97	2.26	0.83	0.17
38	531	176	0.91	46.28	2.36	0.83	0.17
39	538	174	0.92	48.68	2.46	0.84	0.16
40	545	172	0.92	51.20	2.57	0.84	0.16
41	552	171	0.93	53.82	2.68	0.85	0.15
42	559	169	0.93	56.56	2.80	0.86	0.14
43	566	167	0.94	59.42	2.92	0.86	0.14
44	574	166	0.94	62.39	3.04	0.87	0.13
45	581	164	0.95	65.50	3.17	0.87	0.13

Table A.4 Conversion factors

Length	1 m = 10^2 cm
Area	1 m^2 = 10^4 cm^2
Volume	1 m^3 = 10^6 cm^3
Density	1 kg m^{-3} = 10^{-3} g cm^{-3}
Pressure	1 pascal = 10^{-2} mbar
Heat	1 Joule = 0.2388 cal
Heat flux	1 Watt = 0.8598 kcal hr^{-1}
Heat flux density	1 W m^{-2} = 0.8598 kcal $m^{-2}hr^{-1}$
	1 W m^{-2} = 1.433 × 10^{-3} cal cm^{-2} min^{-1}
	1 W m^{-2} = 2.388 × 10^{-5} cal cm^{-2} s^{-1}
Heat transfer coefficient	1 W $m^{-2}K^{-1}$ = 0.8598 kcal m^{-2} hr^{-1} K^{-1}
Specific heat	1 J $kg^{-1}K^{-1}$ = 0.2388 cal $kg^{-1}K^{-1}$
Thermal conductivity	1 W m^{-1} K^{-1} = 0.8598 kcal $m^{-1}hr^{-1}$ K^{-1}
	1 W $m^{-1}K^{-1}$ = 2.388 × 10^{-3} cal cm^{-1} s^{-1} K^{-1}
Thermal resistance	1 s m^{-1} = 10^{-2} s cm^{-1}
Force	1 Newton = 10^5 dynes
Work, energy	1 Joule = 10^7 ergs
Power	1 Watt = 10^7 erg s^{-1}

Table A.5 Physical constants

Constant	Value	
Speed of light in vacuum	2.997925 × 10^8	m s^{-1}
Avogadro constant	6.02252 × 10^{23}	$mole^{-1}$
Planck constant	6.6256 × 10^{-34}	J s
Gas constant	8.3143	J $mole^{-1}K^{-1}$
Boltzmann constant	1.38054 × 10^{-23}	J K^{-1}
Stefan-Boltzmann constant	5.6697 × 10^{-8}	W $m^{-2}K^{-4}$

References

A.1 Calder, W. A. and J. R. King (1974) Thermal and caloric relations of birds. *Avian Biology,* Vol. 4 (D. S. Farner, J. R. King, eds.). New York: Academic Press.

A.2 Monteith, J. L. (1973) *Principles of Environmental Physics.* New York: American Elsevier.

A.3 Robinson, D. E., G. S. Campbell, and J.R. King (1976) An evaluation of heat exchange in small birds. *J. Comp. Physiol. 105*:153–166.

A.4 Winslow, C. E. A., L. P. Herrington, and A. P. Gagge (1937) Physiological reactions of the human body to varying environmental temperatures. *Amer. J. Physiol. 120*:1–22.

Subject Index

A

B

C

G

H

I

K, L

M

N

O

P

R